Mathematics
in Ancient Egypt

Mathematics in Ancient Egypt

A Contextual History

Annette Imhausen

Princeton University Press
Princeton and Oxford

Requests for permission to reproduce material from this work should be sent to Permissions, Princeton University Press
Published by Princeton University Press, 41 William Street, Princeton, New Jersey 08540
In the United Kingdom: Princeton University Press, 6 Oxford Street, Woodstock, Oxfordshire OX20 1TR
press.princeton.edu

Cover image: Model granary from tomb of Meketre. Middle Kingdom, ca. 1981–1975 B.C. Wood, plaster, paint, linen, grain, H. 36.5 (14 3/8 in.); l. 74 (29 1/8 in.); w. 58 cm (22 13/16 in.). Egyptian; Thebes, Meketre. Rogers Fund and Edward S. Harkness Gift, 1920 (20.3.11). © The Metropolitan Museum of Art. Image source: Art Resource, NY

First paperback printing, 2020
Paperback ISBN 978-0-691-20907-4

The Library of Congress has cataloged the cloth edition as follows:
Imhausen, Annette.
    Mathematics in ancient Egypt : a contextual history / Annette Imhausen.
        pages cm
    Includes bibliographical references and index.
    ISBN 978-0-691-11713-3 (hardcover : alk. paper) 1. Mathematics, Egyptian. 2. Mathematics—History. I. Title.
    QA27.E3I43 2015
    510.932—dc23

                                                                          2015009708

British Library Cataloging-in-Publication Data is available

This book has been composed in Minion Pro

Printed in the United States of America

Contents

NEW KINGDOM   127

GRECO-ROMAN PERIODS   179

Preface

This book is the result of the first ten years of my postdoctoral research. Despite the length of time it took to complete it, I am very aware that the results described in the book are temporary at best. Much work still has to be done in all areas of Egyptian mathematics, from its very beginnings, i.e., the development of numbers and metrological systems, until its final stages in Demotic or even Coptic Egypt. But even those areas that have traditionally been the focus of research on Egyptian mathematics, i.e., the mathematical papyri from the Middle Kingdom/Second Intermediate Period, are not fully explored. However, despite all these aspects of non-definitiveness, I hope that this book will still be able to serve as an introduction to Egyptian mathematics for those who are interested in it. The essential place that mathematics held within the ancient Egyptian culture is also reflected by its direct or indirect occurrence in texts other than mathematical and administrative documents. I have included examples of these texts and in doing so have drawn freely on the work of many scholars. Translating ancient Egyptian texts is never a trivial task, and as time has shown, translations will alter with our growing knowledge of the Egyptian language and culture. The translations given in this book thus represent the authors' understanding at the time the translation was done.

In the years that this book was conceived and written, I have experienced much support, which I am grateful to acknowledge. Even after the completion of my dissertation project and subsequent postdoctoral fellowships, my former Doktorväter Jim Ritter and David Rowe have been invaluable as teachers, mentors, colleagues, and friends and I am especially grateful to both of them. Work on this book was begun during a Junior research fellowship at Cambridge, UK, and I am very grateful to Trinity Hall, Cambridge, and its then-master Peter Clarke for providing fantastic surroundings for work and life. Likewise, the fellows and staff of Trinity Hall, most notably Colin Austin and Tom Körner, made the time in England something that I will never forget. Life in Cambridge would not have been the same without the friendly reception that I had at the department of Oriental Studies and the Department of History and Philosophy of Science. I am grateful to John Ray, Eleanor Robson, Kate Spence, and Liba Taub.

During my work on this book I have experienced many pleasant academic encounters, which in some cases have turned into friendships, for which I am grateful. Knowing my limited ability for completeness, I refrain from adding a list of names here with one exception whom I would like to thank in lieu of all the aforementioned colleagues and friends: Kim Plofker and I shared an office at the Dibner Institute, and in this case officemates became friends. Since we left the Dibner, we have visited each other on a regular (if longish interval) basis, and I will not give up hope to share an office with her again at some point in the future. For the support that I have experienced in various academic institutions, most notably libraries and museums, I would like to thank their respective staffs who have made research a pleasant experience. Richard Parkinson and Stephen Quirke who have since moved on to further academic positions kindly provided access and information at various stages of research.

After several years of research and teaching in various interesting places, the history department at Frankfurt University has provided me with a permanent position. I am grateful to my colleagues, especially Moritz Epple, for their friendly welcome in the department, which provides a good environment for teaching and creative research. I also must thank my team at Frankfurt, Susanne Bernhart and Daliah Bawanypeck, who are of invaluable support. Our student assistant, Nadine Eikelschulte, has drawn some of the figures for the book.

Finally, I would like to express my gratitude for the financial support that I have had over the years, which—as someone working on a subject as esoteric as ancient Egyptian mathematics—I never considered could be taken for granted and which therefore always came as a pleasant surprise. Two grants have been of special importance for this book: The Gerda Henkel Stiftung granted the money to enable me to begin working on Demotic mathematics. During a sabbatical provided by the Frankfurt Cluster "Formation of Normative Orders," the manuscript of the book was reworked and finally submitted to Princeton in 2014.

The book was intended to be published by Princeton University Press from the beginnings of its conception. I would like to thank the various people involved with this project, most notably Vickie Kearn, who accompanied the project from the beginning, as well as Quinn Fusting, Leslie Grundfest, and Linda Thompson, who were a pleasure to work with in getting from manuscript to book. I would like to acknowledge that it is a pleasure to have a manuscript professionally copyedited, which has improved every page. Any remaining errors and misunderstandings are entirely mine.

Finally, my life does not consist of scholarship alone. I met my friend Simone Weller while still at university, and we plus our since-added family members have never lost touch. We are godparents to one another's daughters and have thus moved from friendship to family.

My life would not have been the same without her. My husband Paul has been part of my life since the beginnings of the book. We have lived together in three countries and some years ago embarked on the adventures of parenthood with our two girls, Emma and Sophia. For his support over all this time (and looking foward to the coming years . . .) this book is dedicated to him.

Mathematics
in Ancient Egypt

INTRODUCTION

Ancient Egypt has left us with impressive remains of an early civilization. These remains also directly and indirectly document the development and use of a mathematical culture—without which, one might argue, other highlights of ancient Egyptian culture would not have been possible. Egypt's climate and geographic situation have enabled the survival of written evidence of this mathematical culture from more than 3000 years ago, so that we can study them today. The aims of this book are to follow the development of this early mathematical culture, beginning with the invention of its number notation, to introduce a modern reader to the variety of sources (often, but not always, textual), and to outline the mathematical practices that were developed and used in ancient Egypt. The history of ancient Egypt covers a time span of more than 2000 years, and although changes occurred at a slower pace than in modern societies, we must consider the possibility of significant change when faced with a period of this length. Consequently, this book is organized chronologically, beginning around the time of the unification of Egypt around 3000 BCE and ending with the Greco-Roman Periods, by which time Egypt had become a multicultural society, in which Alexandria constituted one of the intellectual centers of the ancient world.

Each section about a particular period analyzes individual aspects of Egyptian mathematics that are especially prominent in the available sources of this time. Although some of the features may be valid during other periods as well, this cannot simply be taken for granted and is not claimed. Covering a time span this large also means that the material presented can be only a selection of all possible available sources. Similar to other areas of Egyptian culture, the source situation is often problematic. As we will see later, the most detailed sources for Egyptian mathematics, Egyptian mathematical texts, are available only from the time of the Middle Kingdom (2055–1650 BCE)[1] onward, and even then we have only about half a dozen

---

1   Dates given are taken from Shaw, *History*.

chance finds, which cover a period of about 200 years. From the New Kingdom (1550–1069 BCE), practically no mathematical texts have survived. It is only in the Greco-Roman Periods (332 BCE–395 CE) that a second group of mathematical texts of approximately the same quantity as the earlier Middle Kingdom material is extant. However, not only mathematical texts themselves inform us about Egyptian mathematics. If the available sources in the form of architectural drawings, administrative documents, literary texts, and others are taken into account, a much more complete picture of Egyptian mathematical culture becomes possible. I have chosen those sources that I hope to be the most significant in illustrating the mathematics of ancient Egypt.

The following parts of this introduction shall serve to provide the reader with a background in the historiography of Egyptian mathematics and the problems and possible approaches in the historiography of ancient mathematics (technical aspects vs. contextual, sociological, and cultural aspects), as well as indicate specific difficulties inherent in Egyptian sources.

0.1 PAST HISTORIOGRAPHY

The study of Egyptian mathematics has captured the interest of modern scholars for almost 200 years. Some try to find the foundations of the impressive buildings—still evident today in remains of pyramids, temples, and tombs all over Egypt—which were built under the auspices of Egyptian pharaohs. Others are drawn to the subject by a fascination with mathematics and its earliest foundations: Egypt and Mesopotamia were the first cultures to develop sophisticated mathematical systems, and these have appealed to mathematicians as well as math students interested in the history of their subject. The way these were organized, their distinct characteristic features, which in some respects differ greatly from our modern mathematical habits while being surprisingly similar to what we do today in others, have sparked the fascination of many.

But it is also relevant—and due to the extant sources maybe most interesting of all—to the historians or Egyptologists who study practical aspects of daily life in ancient Egypt. Then, as today, mathematics was needed in daily life.

Consequently, the mathematical system developed in pharaonic Egypt was practically oriented, designed to satisfy the needs of bureaucracy. Therefore, it is "more than mathematics" that we can find in Egyptian mathematical sources.[2] By reading mathematical texts, we

---

2 See Robson, "More than metrology," p. 361: "It turns out that a wealth of interesting insights can be gained from mathematical material that has traditionally been dismissed as unimportant and trivial."

can learn about the practical backgrounds that required and shaped the evolving mathematical knowledge. They inform us about the exchange of bread and beer, the distribution of rations, work rates of different professions, and other aspects of daily life. Thereby the information found in mathematical texts complements the evidence from administrative documents and archaeological finds.[3]

The first publications about mathematics in pharaonic Egypt appeared at the second half of the nineteenth century, after the now-famous Rhind mathematical papyrus had made its way into the collections of the British Museum and thereby made itself available for study.[4] Even today, the Rhind mathematical papyrus constitutes our most important source, and its initial publication in 1877 was followed by two further editions in 1923 and 1927, as well as numerous articles and some monographs on Egyptian mathematics.[5] The 1877 edition of the Rhind mathematical papyrus was followed in 1898 by the publication of the major mathematical fragments from the Lahun papyri and in 1900 and 1902 by the publication of two fragments of a mathematical papyrus (papyrus Berlin 6619) kept in Berlin.[6] The volume of the *Catalogue général des antiquités égyptiennes du Musée du Caire* of the ostraca published by Georges Daressy in 1901 also included two wooden boards of mathematical content. Daressy published his interpretation in 1906, which was corrected in 1923 by Thomas Eric Peet.[7] The mathematical leather roll (BM 10250), which had arrived at the British Museum together with the Rhind mathematical papyrus, was too brittle to be unrolled immediately. After a method for dealing with this brittle leather was developed, it was unrolled in 1926 and was published in the following year by Stephen R. K. Glanville.[8] The second major source, the Moscow mathematical papyrus, was not published until 1930.[9] Since then, only two further mathematical texts have been found, both of them ostraca with only few (incomplete) lines of text.[10]

3   This was used, for example, by Dina Faltings in her work on ancient Egyptian baking and brewing (Faltings, *Lebensmittelproduktion*).
4   Even before the Rhind papyrus was available, publications on Egyptian numbers and arithmetic based on the inscriptions of the Edfu temple appeared. See Brugsch, "Rechenexempel."
5   Eisenlohr, *Mathematisches Handbuch*; Peet, *Rhind Mathematical Papyrus*; Chace, Bull, Manning, and Archibald, *Rhind Mathematical Papyrus*. Following the controversial publication of Eisenlohr, the British Museum published its own facsimile in 1898. On this controversy and the quality of the British Museum facsimile, see Imhausen, *Algorithmen*, p. 8, with notes 11 and 12.
6   Lahun fragments: Griffith, *Petrie Papyri*, pp. 15–18. Griffith included only those fragments in his publication that seemed well-enough preserved; therefore, some of the material (including some mathematical fragments) remained unpublished until the complete edition of all Lahun papyri in 2002–2006 (Collier/Quirke, *UCL Lahun papyri*), Vol. 2, pp. 71–96. Berlin 6619 fragments: Schack-Schackenburg, "Berlin Papyrus 6619," and Schack-Schackenburg, "Kleineres Fragment."
7   Daressy, *Catalogue général*; Daressy, "Calculs égyptiens"; Peet, "Arithmetic."
8   Glanville, "Mathematical leather roll." The last part of this publication (pp. 238–39) describes the method used to deal with brittle leather. See also Scott and Hall, "Laboratory notes."
9   Struve, *Mathematischer Papyrus Moskau*.
10 Hayes, *Senmut*, No. 153, and López, *Ostraca Ieratici*, No. 57170.

Two monographs on Egyptian mathematics have been published during the twentieth century. In 1972, Richard Gillings, a historian of mathematics based in Australia, published his *Mathematics in the Time of the Pharaohs*, which is still available from Dover publishers.[11] In 1993, the French monograph *Mathématiques égyptiennes*, by Sylvia Couchoud, was published.[12] Two further book-length studies are the dissertation and habilitation of the German Egyptologist Walter Friedrich Reineke: *Die mathematischen Texte der Alten Ägypter* (Berlin 1965) and *Gedanken und Materialien zur Frühgeschichte der Mathematik in Ägypten* (Berlin 1986). While the dissertation remains unpublished (although photocopies are held in the libraries of some German Egyptological institutes), the habilitation was published in 2014, almost 30 years after its completion. In 2014, a French publication on Egyptian mathematical texts appeared, which—despite its author's knowledge of recent publications on Egyptian mathematics—reverts to writing an assessment of Egyptian mathematics from a modern mathematician's point of view.[13] In addition, a source book of Egyptian mathematics was published in 1999.[14]

Much more numerous, in fact, too numerous even to be listed completely, are works on individual aspects of Egyptian mathematics, mostly in the form of articles in Egyptology and history of mathematics journals, but some also full monographs. Due to the large number of studies, only selected works will briefly be sketched in the following paragraph. An aspect of Egyptian mathematics that has fascinated historians of mathematics is the Egyptian method of fraction reckoning. Expressed in modern mathematical terminology, ancient Egyptian mathematics used only unit fractions, with the only exception being the fraction $\frac{2}{3}$. The first two researchers to publish monographs on aspects of Egyptian fraction reckoning were Otto Neugebauer and Kurt Vogel, who both wrote their doctoral theses on this subject.[15] A central part in all investigations of Egyptian fraction reckoning was the composition of the so-called $2 \div n$ table, a table that must have been essential in handling fractions during standard computations. From a mathematical point of view, the representation of a fraction as a set of unit fractions is not unambiguous; however, the two extant copies of the $2 \div n$ table indicate that a standard representation was used in ancient Egyptian mathematics. This led to the question of how the individual entries were chosen. A controversial contribution was published by Richard Gillings, who tried to establish a set of rules that were used when the table

11   Gillings, *Mathematics in the Time of the Pharaohs*.
12   Couchoud, *Mathématiques égyptiennes*.
13   Michel, *Mathématiques de l'Égypte ancienne*.
14   Clagett, *Egyptian Mathematics*. See the reviews Allen, "Review Clagett," and Spalinger, "Review Clagett".
15   Neugebauer, *Bruchrechnung*, and Vogel, *Grundlagen der ägyptischen Arithmetik*. Other contributions on Egyptian fraction reckoning (in chronological order) are Rising, "Egyptian use of unit fractions," Bruins, "The part," Bruins, "Reducible and trivial decompositions," Rees, "Egyptian fractions," and Knorr, "Techniques of fractions."

was compiled.[16] However, although these rules explain some of the choices, overall it is not possible to predict the answer that is found in the table based on these rules; therefore, mathematicians keep coming back to work on its analysis.[17] The modern description of Egyptian fraction reckoning as being "restricted" to unit fractions is obviously anachronistic (indeed, the Egyptian concept of fractions did not include a numerator, but from a historian's point of view this cannot be criticized on the basis that our modern fractions consist of denominator and numerator). Furthermore, this criticism does not do justice to the development of Egyptian fractions. Finally, the representation of Egyptian fractions in our modern system using the numerator 1 throughout causes Egyptian fraction reckoning to look more cumbersome than necessary to a modern reader, which has led to negative assessments of modern researchers, as for example that of Otto Neugebauer:

> The primitive, strictly additive, Egyptian way of computing with unit fractions had a detrimental effect throughout, even on Greek astronomy.[18]

Fraction reckoning was, without any doubt, a demanding part of Egyptian mathematics; the fact that all the extant Egyptian tables are for fraction reckoning (either for absolute numbers or for metrological systems) bears testimony to the inherent intricacies of this area of Egyptian mathematics. However, the Egyptian system also apparently had its advantages, and the available tables presumably reduced the effort that was needed. It is remarkable that Egyptian fractions prove to be resistant against the contact with Mesopotamian mathematics (which used sexagesimal fractions) and that Egyptian fractions also appear in Greek mathematics as well as the Western mathematics of the Middle Ages.[19]

Other aspects on which researchers have focused are the Egyptian calculation of the area of a circle,[20] the way in which the Egyptian method to calculate the volume of a truncated pyramid was obtained,[21] and the method of solution for a set of problems that—again phrased in anachronistic modern terminology—are similar to our algebraic equations.[22]

---

16 Gillings, "Divisions of 2," and Gillings, *Mathematics in the Time of the Pharaohs*, pp. 45–80. His results were questioned in Bruckheimer and Salomon, "Some comments" and defended by Gillings in, "Response."
17 For example, van der Waerden, "(2:*n*) table" and the recent contribution by Abdulaziz, "Egyptian method."
18 Neugebauer, *History of Ancient Mathematical Astronomy*, p. 559.
19 See, for example, Fibonacci's Liber Abaci from the early thirteenth century (Sigler, *Fibonacci-Liber Abaci*).
20 Gillings and Rigg, "Area of a circle," Smeur, "Value equivalent to π," Engels, "Quadrature of the circle," and Gerdes, "Three alternate methods."
21 Gunn and Peet, "Four geometrical problems," Vogel, "Truncated pyramid," Struve, *Mathematischer Papyrus Moskau*, pp. 174–76, Thomas, "Moscow mathematical papyrus, no. 14," Luckey, "Rauminhalt," Vetter, "Problem 14," Gillings, "Volume of a truncated pyramid," Neugebauer, "Pyramidenstumpf-Volumen."
22 For a discussion, see Imhausen, "ꜥḥꜥ-Aufgaben."

The amount of available literature on Egyptian mathematics is even more astonishing if we take into account that the only sources on which almost all these studies were founded are four papyri (of which half consist only of a number of fragments), a leather roll, a wooden board, and two ostraca (stone or pottery shards used as writing material)—almost all from the Middle Kingdom (2055–1650 BCE), with five more papyri and some further ostraca—from the Ptolemaic and Roman Periods (332 BCE–395 CE), more than a thousand years later. The Ptolemaic and Roman sources (the number of which is likely to increase from unpublished texts in museums as well as possible new finds) were rarely taken into consideration. Studies of Egyptian mathematics concentrated mostly on the earlier material. In spite of all these limitations, the sources still allow us new insights and were sufficient material for many studies over the past 125 years. In fact, we seem far from exhausting these sources. Due to developments in the history of mathematics of the last 40 years, it has now become obvious that many "statements" about Egyptian mathematics that were made a long time ago and that have since then been accepted as "truths" need to be reassessed.[23] At the beginning of the twentieth century, a period in which research on Egyptian mathematics reached its first boom, questions of *how* ancient mathematical knowledge related to later developments (e.g., Greek geometry), including our modern mathematical system, determined the research on ancient sources. In the respective works, the unfamiliar ways in which mathematical operations were expressed were translated into their "modern equivalents." The insight that mathematics is not culturally independent, that it is not constantly evolving in a linear way toward the next level, was gained much later—and this insight and its consequences are still debated.[24] As a consequence, "close reading" of ancient mathematical sources evolved, in which the technical language and the formal structures of the original texts are taken seriously.[25] As a further consequence, ancient mathematics was understood as best studied in relation to the culture in which it evolved.[26]

---

23 For an overview of the most common "myths" in the historiography of Egyptian mathematics, see Imhausen, "Myths."

24 One of the essential articles was Unguru, "Need to Rewrite," which was elaborated in Unguru and Rowe, "Quadratic Equation I" and Unguru and Rowe, "Quadratic Equation II." The importance of the discussion that was sparked by Sabetai Unguru has been acknowledged in the volume Christianidis, *Classics*, which includes a reprint of Unguru, "Need to Rewrite" and its first responses. Twenty-five years later, Sabetai Unguru and Michael Fried have added another monograph to the discussion (Fried and Unguru, *Conica*).

25 This has affected Egyptian mathematics as well as Mesopotamian mathematics. Fundamental in this respect for Mesopotamian mathematics are the works of Jens Høyrup, Jim Ritter, and Eleanor Robson; see, for example, Høyrup, *Lengths, Widths, Surfaces*; Ritter, "Reading Strasbourg 368," and Robson, *Mathematics in Ancient Iraq*.

26 See, for example, Imhausen, "Egyptian mathematical texts and their contexts." Again, this is also true for other areas of ancient mathematics; for Mesopotamian mathematics, see Robson, *Mathematics in Ancient Iraq*; for Greek and Roman mathematics, Cuomo, *Ancient Mathematics*.

At the same time, however, traditional work on Egyptian mathematics continues, as does the publication of unfounded speculation about Egyptian mathematics (or even Egyptian science in general), which also had a recent boom with the Internet as a platform, where this kind of material can be made easily available. The scarce source material has not been, and presumably never will be, able to answer every question asked in modern times. This has encouraged rather speculative theories founded on practically no evidence; in fact, it is exactly the lack of evidence that has enabled speculations of this kind. Classical examples are the methods that were used to align and build the pyramids as well as the way Egyptian mathematical knowledge was discovered.[27] The honest and academically responsible answer to most of these questions (at least at the present moment) would simply be that we don't know. For some of them, at least, possibilities may be discussed, while others may almost certainly never be answered. With the same certainty, however, there will always be some trying to fill these gaps with their individual ideas. In assessing this kind of speculation, the actual available evidence must remain in the picture. Therefore, by presenting a variety of available sources and pointing out the limitations of the available material, one aim of this book is to encourage its readers to judge speculations about Egyptian mathematics with a critical and informed eye.

The progress in the history of mathematics, which has influenced recent studies of Egyptian mathematics, has run parallel to an improvement in our understanding of Egyptian languages and culture. Not many of the mathematical texts, however, have yet benefitted from this development. Color photographs of the Rhind mathematical papyrus, along with a brief overview of Egyptian mathematics, were published in 1987 by Gay Robins and Charles Shute.[28] The third-largest source is the Lahun mathematical fragments, which Jim Ritter and I have reedited in the context of the new edition of the Lahun papyri.[29] It is to be hoped that other sources will follow.

0.2 AIMS OF THIS STUDY

Traditionally, the mathematical texts, especially the hieratic mathematical texts, have been the main sources in works on ancient Egyptian mathematics. There are obvious reasons for this. The mathematical texts indicate specific mathematical problems and also detail how they were solved, giving the most immediate access to ancient mathematical techniques and concepts.

---

27   For a detailed discussion of the theories and their shortcomings of the architecture of the pyramids and the mathematics involved, see Rossi, *Architecture and Mathematics*.
28   Robins and Shute, *Rhind Mathematical Papyrus*.
29   Imhausen and Ritter, "Mathematical fragments," and Imhausen, "UC32107A."

However, as Jim Ritter has elaborated, the problems that a historian of ancient mathematics faces in the interpretation of these texts are many and varied:

> What does a text mean? A question like this poses particularly acute problems for someone who works on ancient texts. This is not to underplay the complexity and plurality of interpretations in, say, contemporary literary texts, but rather that in the case of Antiquity we are often confronted with the opposite situation: the difficulty of finding even a single possible interpretation. The problem may occur at a very concrete level: lacunae in the texts, *hapaxes*, a technical vocabulary for which it is difficult or impossible to fix the semantic referents. Moreover, we are often confronted with contemporary constraints, not always immediately perceptible to us, concerning, among other things, the rules of the textual genre, the available concepts or techniques of expression, the aims pursued by the authors.[30]

These difficulties are reflected, at least in part, in the amount of available literature on the Egyptian mathematical texts. In this book on Egyptian mathematics, the mathematical texts obviously remain important sources. Following recent developments in the methodology of ancient mathematics, I will try to describe the characteristics of Egyptian concepts and techniques and especially focus on the differences from our modern concepts. Fundamental in this respect is to realize that algebraic equations, an extremely powerful mathematical tool, did not exist in ancient Egypt. Instead, the mathematical solutions are indicated in the form of procedures, step-by-step instructions that—if followed—will lead to the solution of the given problem.

The aim of this study is to go beyond the few hundred years covered by the still available mathematical texts. The beginnings of Egyptian mathematics, in the form of number notation and metrological systems, can be traced much further back if other sources are taken into account. I will attempt to sketch the development of Egyptian mathematics from the invention of number notation, which occurred at approximately the same time as the invention of writing, until the Greco-Roman Periods using a variety of available sources, thereby also describing the context and cultural setting of Egyptian mathematics throughout pharaonic history. In order to achieve this, archaeological sources (such as drawings on the walls of tombs), administrative texts, autobiographies, and various literary texts will be used. Obviously, information

---

30   Ritter, "Reading Strasbourg 368," p. 177.

to be received about specific mathematical techniques from these sources is very limited indeed; however, they are extremely useful in assessing the role of mathematics within pharaonic culture and in getting glimpses at those people who may have been the authors of the mathematical texts that we have been studying in isolation for the past 150 years.

A chronological order was chosen to help the reader place the individual evidence within its respective time frame. Egyptian history spans a period of more than 3000 years, and it has to be kept in mind that individual developments did take place during that time. It is not legitimate to simply mix evidence from different stages of Egyptian history; however, at least at some times and instances it seems plausible to assume a certain continuity even in the absence of direct evidence. To provide a historical frame, each section will include a very brief sketch of historical and political events for that period.[31]

---

31 A more detailed overview and introduction to the respective times can be found at Shaw, *History*. For an overall introduction into ancient Egyptian history and culture, see the revised British Museum book of ancient Egypt (Spencer, *Ancient Egypt*).

PREHISTORIC AND EARLY DYNASTIC PERIOD

The Predynastic and Early Dynastic Periods enclose the time of Egypt's transition from multiple predynastic settlements into a unified state, which was centrally governed and administered under a single king.[1] This transition was presumably initiated by the centers that evolved in several locations in Egypt. Predynastic centers were Maadi (ca. 3900–3500 BCE) and Buto (ca. 3500–3200 BCE) in Lower Egypt and Badari (ca. 4500–4000 BCE) and Naqada (subdivided into Naqada I, ca. 4000–3500 BCE, Naqada II, ca. 3500–3200 BCE, and Naqada III, ca. 3200–3000 BCE) in Upper Egypt.[2]

The available evidence for both parts of Egypt varies distinctively. While evidence for predynastic cultures in Lower Egypt "consists mainly of settlements with very simple burials in cemeteries," the evidence of predynastic cultures in Upper Egypt includes few settlement data but "cemeteries with elaborate burials," which "represent the acquired wealth of higher social strata."[3] These Upper Egyptian cemeteries have allowed researchers to trace a developing social stratification, with an elite "who controlled the trade in raw materials and organized their transformation into profitable luxury items."[4] Using the spread of ceramics as an indicator for the expansion of the Naqada culture, even the northern delta had been reached by the end of the Predynastic Period.

---

1   For detailed accounts of this part of Egyptian history, see Bard, *Farmers to Pharaohs*, Wilkinson, *State Formation*, and Midant-Reynes, *Prehistory*, especially pp. 167–250. For more compact overviews, see Bard, "Egyptian predynastic," Bard, "Emergence of the Egyptian state," or Adams and Cialowicz, *Protodynastic Egypt*. See also Reineke, *Gedanken und Materialien*, pp. 18–25.

2   Dates after Bard, *Archaeology of Ancient Egypt*. The dating is based on changing pottery styles and was initially devised by William Flinders Petrie at the beginning of the twentieth century. It has since been refined, for example, by Werner Kaiser, Barry Kemp, Stan Hendrickx, and Toby Wilkinson.

3   Bard, "Egyptian predynastic," p. 280.

4   Midant-Reynes, *Prehistory*, p. 237.

How exactly the elite emerging at Naqada managed to enlarge their territory over all of Egypt is likely to remain a matter of scholarly debate, since the evidence supports several possible developments as Béatrix Midant-Reynes outlines:

> Was the process peaceful or warlike? The Narmer Palette—one of the first pieces of written information concerning Egyptian history—certainly implies a military process, apparently showing the southern king conquering the north. Indeed, this air of violence, which constantly appears in the protodynastic evidence, had already made a discrete appearance even earlier, in the so-called Painted Tomb at Hierakonpolis. However, there is no archaeological evidence to support this theory. Dieter Wildung notes that the grave goods in the cemetery of Minshat Abu Omar are much more evocative of a peaceful trading community than a group of warriors. The unification seems to have been not so much an act of conquest as a process of continuous change.[5]

Undisputed seems to be that the southern culture dominated the northern cultures and is at the head of the newly formed unified country. The earliest extant written sources from Egypt originate around the time of the unification. The evidence for the invention of script (and also number notation) is closely linked to Abydos and its cemetery Umm el-Qaab. Abydos, located approximately 500 km south of Cairo, was a major cultural center from predynastic times on. Three areas are distinguished within the cemetery Umm el-Qaab: cemetery U (with approximately 650 tombs dating to Naqada I-III), cemetery B (tombs of the last predynastic and first dynastic rulers), and several large tomb complexes of the First and Second Dynasties.[6] The evolution of mathematics in ancient Egypt begins with the evidence from these cemeteries, especially that from the tomb U-j, which includes the earliest written evidence that has been found.

During the end of the Naqada III period, Egypt was finally unified under a single king who—based on the evidence of grave types, pottery and artifacts—probably descended from the Naqada culture of Upper Egypt. The consolidation of the new state was strengthened by writing in several ways: Potmarks and royal seals were used to identify goods and thus facilitate administration. Papyrus rolls, which were to become the main writing material, probably

---

5   Ibid., pp. 281–82.
6   Work at Abydos is still ongoing. On cemetery U see Dreyer, *Umm el-Qaab I,* and Dreyer, *Umm el-Qaab II;* on the later tombs, see Dreyer, "The tombs of the first and second dynasties."

existed by the middle of the First Dynasty. The oldest extant writings on papyrus (dating from the late Fourth and Fifth Dynasties) are administrative texts that display a developed formal arrangement of data.[7] The usefulness of a numeric system in this context, where numbers and quantities of certain goods need to be recorded, is obvious.

In addition to this utilitarian aspect of the use of early writing, there is also a representational one. Depictions of the king that symbolize his legitimacy to rule are accompanied by short inscriptions. Interestingly, these inscriptions sometimes also make use of the number system. Examples of these objects include palettes and maceheads—common to the Egyptian repertoire of artifacts since predynastic times—which are elevated to cult objects by means of elaborate decoration.

---

7   Baines, "Writing: Invention and early development," p. 884.

1.

The Invention of Writing and Number Notation

At the beginning of any development of more complicated or advanced mathematical techniques stands the invention of scripts and number systems—at least in those instances where (written) evidence of mathematics is extant.[1] In Mesopotamia, the origins of literacy and numeracy have been shown to be closely linked and to result from the needs of accounting.[2] Although some of the evidence from Egypt supports a similar claim, the situation seems to be more complex as far as the uses of writing and numeracy are concerned. The source material available for tracing the invention and early development of writing comes, as is often the case with ancient Egyptian artifacts, on the one hand from funerary contexts (tombs) and on the other hand from temples; evidence from early settlements, where these developments presumably at least partly took place, is lacking. The funerary context of the evidence is presumably due to the body of source material in ancient Egypt. The deserts of Egypt, where tombs and temples were located, provided excellent conditions for the preservation of artifacts; hence, practically all the ancient papyri (not only Egyptian but also Greek) originate from Egypt. However, settlements and towns, the places where life and—presumably—the invention and further development of writing happened, were located then as now in the proximity of the Nile, which provided the necessary water. Therefore, ancient Egyptian settlements are often buried under their modern successors and cannot be excavated; in addition, their proximity to water renders unlikely the survival of any organic artifacts from periods several millennia before our present time.

---

1 For a collection of accounts of the invention of script in various cultures, see Houston, *First Writing*.
2 See Cooper, "Babylonian beginnings," and Robson, "Literacy."

It is estimated that writing was invented in Egypt at the end of the fourth millennium BCE. Among the predynastic elite tombs, tomb U-j (assigned to King Scorpion around 3200 BCE) holds a significant place. Within the twelve rooms of the structure, the earliest evidence of hieroglyphic writing from ancient Egypt was discovered.[3] Two types of objects with inscriptions were found in tomb U-j. On the one hand, almost 200 labels made from bone, ivory, or—exceptionally—stone and on the other hand, ceramic vessels. The labels are mostly approximately 1.5 by 2 cm and originate from the tomb U-j and its proximity. They all include a hole, which presumably served to attach them to objects that have since perished.[4] This type of object is also known from later times (since King Narmer), where it is often used to record deliveries of oil.[5] Similar labels have also been found at Naqada.[6] The ceramic vessels contribute approximately 125 inscriptions, which can be assigned to several groups, among which the group with scorpions (60 attestations) is the largest.[7] While the inscriptions on the ceramic vessels consist of one or two large pictorial signs (e.g., a scorpion), inscriptions on the labels can be divided into two groups, either pictorial signs or (combinations of) abstract signs, which are interpreted as representations of numbers.[8] Based on the regularity of the signs, it is assumed that the evidence found in tomb U-j is not the first writing ever achieved in Egypt but rather the product of a previous evolution.

If we presume that some basic assumptions about the earliest written objects from tomb U-j are correct, that is, if the abstract signs were indeed the representation of numbers or quantities and the labels were attached to some goods about which they held information (e.g., the indication of their quantities or their origin or owner) then, like in Mesopotamia, the earliest writing is linked to administrative needs, and the invention of a quantity/numerical notation along with the invention of script is almost a necessary consequence.

However, the situation in Egypt may be more complex. Along with these administrative records, there are also early attestations of Egyptian writing of a representative character, such as, for example, the notation of the name of the king written in a rectangular enclosure called *serekh*. Writing in Egypt, therefore, was not only used by the elite for their administrative needs, but it was also recognized as a tool to represent and display power. For each of these

---

3   The interpretation of these early written sources is extremely difficult. A first assessment of the material is given in Dreyer, *Umm el-Quaab I*, pp. 113–45. Further discussions can be found in Baines, "Earliest Egyptian writing," Baines, "Writing: Invention and early development," Baines, "Writing and kingship," Breyer, "Schriftzeugnisse," Kahl, "Frühe Schriftzeugnisse," Morenz, "Systematisierung der ägyptischen Schrift," and Regulski, "Origin of writing."

4   Dreyer, *Umm el-Qaab I*, pp. 136–37.

5   Dreyer, *Umm el-Qaab I*, p. 137.

6   C. Dreyer, *Umm el-Qaab I*, p. 139 with figure 83a, and Quibell, *Archaic objects*, CG 14101–14106.

7   Ibid., p. 84.

8   Ibid., p. 113.

functions, a preferred writing surface and, along with it, a preferred way of writing developed. While the representation and display of power were well served by hard stone surfaces, into which an immensely durable inscription was chiseled that proved to last extremely well over the millennia, the daily necessities of administration required a writing that could be executed fast. Along with the hieroglyphic writing system, therefore, a second style of writing was developed, which used abbreviated signs that could also be ligatured to make them even more efficient and which was written with a brush and ink onto papyrus or ostraca (pottery or stone shards). This second type of writing is called *hieratic* and shows a much greater variation in its form over time as well as in the comparison of individual scribal hands.[9] On a rather rudimentary level, one can perhaps compare the hieroglyphic writing to our printed script and the hieratic writing to handwriting.

---

9   Ancient Egyptian writing and language spans over 3000 years and has undergone considerable developments during this time. For an overview see Loprieno, *Ancient Egyptian*.

2.

The Egyptian Number System

The number system used in ancient Egypt can be described in modern terminology as a decimal system without positional (place-value) notation. The basis of the Egyptian number system was 10 (hence *decimal* system), but unlike our decimal place-value notation using the ten numerics 0, 1, 2, 3, 4, 5, 6, 7, 8, and 9, in which the absolute value is determined by its position within the number (e.g., in the number 125, the absolute value of 1 is 1×10^2, the absolute value of 2 is 2×10^1, and the absolute value of 5 is 5×10^0), the Egyptian system used individual symbols for each power of 10, as shown in table 1.[1]

Although there is no information about the choice of the individual signs for the respective values, some of them seem plausible choices. The most basic, the simple stroke to represent a unit, is used not only in Egypt but also in a variety of other cultures, possibly originating from marks on a tally stick.[2] The sign for 10 was interpreted by Gardiner as a "hobble for cattle, without the cross bar," presumably because its variant with the cross bar (Gardiner V19) occurs as ideogram or determinative in the Egyptian word *mḏt* ("stable"), the sign without the cross bar is (according to Gardiner) attested as a phonogram in the plural *mḏwt* ("stables").[3] Why this sign was chosen to represent the number 10 can only be speculated—maybe 10 animals tied to this hobble? The sign to represent 100, a measuring rope, possibly points to a measuring rope that was used to measure the length of a field, which may have had a standard length of 100 cubits.[4]

---

1   The only general study about Egyptian numbers is Sethe, *Zahlen und Zahlworte*, from the beginning of the twentieth century, which remains a very useful reference work. Cf. also Reineke, *Gedanken und Materialien*, pp. 25–43.
2   For an overview cf. Menninger, *Number Words*, pp. 223–48.
3   Gardiner, *Egyptian Grammar*, p. 524. Allen, *Middle Egyptian*, p. 445, lists both signs, V19 and V20 as ideograms.
4   This interpretation can be supported through metrological data; one of the units among the area measures is an area of 100 cubit lengths and 1 cubit width. The measuring rope is also an element of some

TABLE 1: Hieroglyphic signs representing numbers

hieroglyphic sign	Gardiner number	absolute value	phonetic
𓏺	Z1	1	w^c
𓎆	V20	10	md
𓍢	V1	100	$\check{s}t$
𓆼	M12	1000	$\underline{h}\beta$
𓂭	D50	10.000	db^c
𓂽	I8	100.000	$\underline{h}fn$
𓁨	C11	1.000.000	$\underline{h}\underline{h}$

Note: The combination of letter and number ("Gardiner number") refers to the sign list in Gardiner, *Egyptian grammar*.

The white lotus plant, the phonogram for $\underline{h}\beta$, is often depicted in Egyptian representations as a plant in a landscape, as an ornament, and often with symbolic meaning in ritual or religious actions.[5] Possibly its frequency (apparently it can still be found in the Egyptian canals today) within the landscape may have resulted in its use as a symbol for the number 1000. The hieroglyph of a finger as 10,000 is plausible only if the Egyptian construction, similar to the modern English construction, used a concept of a "bundle" of thousands, that is, "ten thousand," and hence the idea of 10 fingers, each of value 1000, makes the use of the finger somewhat plausible. The sign for 100,000, a tadpole, probably does not need much explanation for anyone who has ever seen a lake with frog spawn and the multitude of tadpoles developing from it. Finally, the largest of the Egyptian number symbols, representing 1,000,000, is the seated god *Heh*. *Heh* was one of the eight primeval gods, supposed to carry the heaven under the earth. His depiction is often found on temple walls, vases, and jewelry, where he was supposed to grant millions of years of life.[6]

Each of these signs was written as often as required to represent the number, with groupings of individual signs arranged symmetrically.[7] For example, to write the number 845, the

scribal statues; see, for example, those of Senenmut and Penanhor. On the standard length of fields cf. also Reineke, *Gedanken und Materialien*, p. 84.

5   On the white lotus in ancient Egypt, see Manniche, *Egyptian herbal*, pp. 126–27.

6   Hart, *Gods and Goddesses*, p. 66. Reineke, *Gedanken und Materialien*, p. 26 suggests a connection between the usage of *Heh* for one million and the "infinite sky" linked to this god.

7   A discussion of possible arrangements can be found in Sethe, *Zahlen und Zahlworte*, pp. 4–7.

Egyptian scribe would have placed eight times the sign V1 (arranged as a group of four times the sign V1 over another group of four times the sign V1), followed by four times the sign V20 (arranged in pairs of two), followed by five times the sign Z1 (arranged in a group of three placed over a group of two): ⲅⲅⲅⲅⲅⲛⲛ|||.

A special writing was used up to the Sixth Dynasty in hieroglyphic representations of numbers for indicating multiple thousands. Several lotus flowers were depicted not as individual groups of signs but rather as a bush of flowers.[8]

This style of number writing did not require a symbol for 0—where our number system employs the 0 to indicate an empty place, the Egyptian notation would simply lack the respective sign altogether. Thus, in the writing of the number 100, our 0 indicating lack of tens and units is indicated in the Egyptian system by the absence of these signs. Therefore, the Egyptian number system does not require a symbol for 0. The modern 0 also fulfills some other functions; for example, it also indicates if a balance has come to 0 or the result if, for instance, 4 is subtracted from 4. In these cases, Egyptian writing used the hieroglyph ⲟ (F35), the phonogram *nfr*.[9] *Nfr* has the basic meaning "good" or even "perfect" and thus may have indicated the lack of a remainder in these cases. *Nfr* is also used as a negative word in combination with verb forms and nominal sentences; hence this may also have been the origin of the usage of *nfr*, indicating the lack of "something."[10] Another symbol used where we might employ 0 was ⲙ (D35), "arms in gesture of negation," the ideogram for *jw*, the relative adjective meaning "which is not" to indicate that something was not there.[11]

The hieratic writing of numbers shows, as does hieratic writing in general, simplified sign forms. It also used a set of regular ligatures to write the combination of groups of identical signs, as required by numbers such as 9, 40, and others.[12] Thus individual ligatures existed for each series 1–9 in the various powers of 10 (1, 2, 3, 4, . . . ; 10, 20, 30, 40, . . .), which were written from the highest to the lowest to represent the respective number.

---

8   Sethe, *Zahlen und Zahlworte*, p. 6.
9   Papyrus Boulaq 18 (Scharff, "Rechnungsbuch," plate 5**).
10   The Berlin digital *Thesaurus Linguae Aegyptiae* lists the meaning "zero" for *nfr* in conjunction with the negative word *nfr* (http://aaew.bbaw.de/tla/index.html; Lemma-Nummer 550123, accessed August 1, 2012). See also Gardiner, *Egyptian Grammar*, section 351. Allen, *Middle Egyptian*, p. 402 (chapter 26.29.3) indicates that the negative word *nfr* is used only in combination with other words and explains the absolute use of *nfr* to indicate a zero remainder as an abbreviation of the word *nfrw*, "depletion" (Allen, *Middle Egyptian*, p. 97 (chapter 9.1)).
11   Attested in the Edfu donation text, see Kurth/Behrmann, *Edfu*, p. 394, and Meeks, Texte des donations, pp. 157ff. See also Allen, *Middle Egyptian*, p. 134 (chapter 12.9).
12   For an overview of these ligatures and their changes over time, compare the respective pages in Möller, *Paläographie*.

According to Sethe, the sign for one million, 𓁨, was no longer in use during the Middle Kingdom, and a multiplicative way of number notation for large numbers was developed, in which a multiplicative factor was put under the sign for larger numbers; for example, in Papyrus UC 32161, the number 925,157 is written as 𓁨𓏤𓏤𓏤𓆼𓆼𓆼𓆼𓆼𓆼𓍢𓎆𓎆𓏭.[13] Multiplicative writings are attested for 100.000 and 10.000 (Middle Kingdom and later), rarely also for 1000 (Bilgai stele, end of Dynasty 19).[14]

---

13 Sethe, *Zahlen und Zahlworte*, pp. 8–10.
14 See Sethe, *Zahlen und Zahlworte*, p. 9, and the references given in the respective sections of Möller, *Paläographie*.

3.

Uses of Numbers and Their Contexts in Predynastic and Early Dynastic Times

The earliest attested Egyptian number recording originates, as was described previously, from the tomb U-j in Abydos. However, the ivory and bone tags that are supposed to carry numbers on them do not yet show the Egyptian number system described in the previous section. There seem to have been only three types of signs used in writing a number: a vertical stroke, a horizontal stroke, and a rope (see figure 1).

Of these, the vertical stroke and the rope have successors in the later evidence of Egyptian number notations in the writing of units and hundreds, respectively. It is not clear if the horizontal stroke is not simply another way of writing units—especially since in some instances, the proper orientation of the label (which renders the strokes as being either horizontal or vertical) is not certain.[1] Their current orientation is assigned to them based on the orientation of grain of the material.[2] Günter Dreyer suggested that the horizontal stroke may be explained as a possible sign for 10 by assuming that the number tags were meant to record quantities of fabric, which were kept in wooden boxes and have since perished.[3]

The individual tags from the tomb U-j do not mix different numerical symbols. Thus we either find vertical strokes or horizontal strokes but never a combination of the two. Note the symmetric grouping of the signs in tags with vertical and horizontal strokes. The hundreds appear in two versions, either with an additional stroke or only as the curled rope. Rather than

---

1   For example, Dreyer, *Umm el-Qaab I*, p. 115, nos. 16 and 17.
2   Dreyer, *Umm el-Qaab I*, p. 139.
3   Evidence from the Old Kingdom supports this interpretation; see Dreyer, *Umm el-Qaab I*, p. 140 with reference to Posener-Kriéger, "Mesures des étoffes," p. 86.

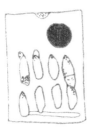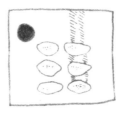

FIGURE 1: Examples of tags with number notations from predynastic tomb U-j (drawn after Dreyer, *Umm el-Qaab I*, Figures 74 and 75)

interpreting the curled rope and the stroke as standing for 100 + 1,[4] it seems more likely that we have here two kinds of metrological systems. Thus the curled rope alone may represent 100 in one system, and the curled rope with the stroke may stand for 100 within another system.

The number of strokes on some tags exceeds 9; for instance, there are examples of 12 strokes on one tag. As John Baines correctly points out, this is not likely to indicate a base other than 10 for the Egyptian number system.[5] However, it may point to the fact that the "full use" of a decimal system is still on its way.

Similar tags were also found in other cemeteries, e.g. at Naqada (see the examples in figure 2). The Naqada tags are inscribed on the front and the back, with one side carrying the number notation. They show the next stage in the development of Egyptian number notation; several signs are combined to form a number on one tag. Again, the signs are symmetrically grouped. The decimal nonpositional system seems to be in place using the signs for units, tens, and hundreds, as they are known from later sources.

The evidence of tomb U-j suggests that the invention of writing occurred during the process of establishing a social hierarchy and the expansion of the Naqada culture. This expansion may have created the need for an organized administration in which writing was a powerful tool to denote ownership and provenance of goods.[6] This tool seems to have been limited to the elite from the beginning. In consequence, this may have led to the use of writing for representational and legitimatizing purposes. Writing thus helped the unification of the state and its consolidation.[7]

---

4   Baines, "Earliest Egyptian writing," p. 157.
5   Ibid.
6   Baines, "Literacy," p. 196.
7   Baines, "Communication and display," p. 472, and Bard, "Origins of Egyptian writing," p. 297.

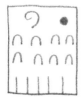 

FIGURE 2: Examples of tags from Naqada (drawn after Quibell, *Archaic objects*, Volume 2, Plate 43)

The context of this early evidence is purely funerary; however, it is probably reasonable to assume that the administration of goods displayed in this context was also common and presumably originated from a daily life practice, which was then taken over into the funerary context.

The Egyptian number system was fully developed at the time of King Narmer, as is documented by his ritual macehead, which was found at the temple of the god Horus at Hierakonpolis, the most important predynastic site in the south of Egypt (cf. figure 3).

The object originated from a ceremonial context, as is indicated not only by its findspot but also by its material, soft limestone. The macehead is carefully decorated with the scenes shown in figure 3, depicting King Narmer, who receives a tribute.[8] King Narmer, the central figure toward whom the scenes are oriented, is shown seated on a throne in a shrine. The shrine is elevated and can be reached via a stair of nine steps. King Narmer is shown wearing the red crown of Lower Egypt, he is dressed in a long robe, and he holds a flail. The area of the throne is protected by a vulture, which is hovering above its roof. Under the king, two fan bearers are depicted (possibly to be understood to have been placed to both sides of the king). Another five figures, further subordinates of the king, are shown in two rows behind the throne. Above them, the *serekh* with the name of the king is shown. In front of the king are three registers with depictions oriented toward the king. In the upper register, four men with standards are shown. The middle register, immediately opposite the seated king, shows another person, who is seated in a litter. Behind the litter, a group of three bearded figures is depicted; the group is framed by two sets of three crescent-shaped signs. The lower register holds the depiction of the tribute, which is presented to King Narmer and which is the earliest representation of the full Egyptian number system, even before the first dynasty. To the right of these three registers is the representation of various animals: an ibis and an enclosure of three antelopes.

---

8   For a description of the mace and its scenes, see Quibell, *Hierakonpolis*, Vol. 2, pp. 39–41.

FIGURE 3: Scenes depicted on the Narmer macehead (drawn after Quibell, *Hierakonpolis*, Volume 1, Plate 26b, by Nadine Eikelschulte)

The scenes on the Narmer macehead, (and, likewise, those on related objects, such as the Scorpion macehead and the Narmer palette) generally represent the civic, priestly, and military functions of the king within the state.[9] Scholars have tried to link these scenes to specific historic events, but our knowledge of this early period does not permit us to confirm any of these claims beyond a doubt. The scenes have been interpreted several times, resulting in interpretations as a concrete commemoration of the unification of Egypt and the foundation of the Egyptian state,[10] as a celebration of a royal festival, as the celebration of a marriage alliance of Narmer, and as the elaborate writing of a year name.[11] For the purpose of this book, it suffices to state that the scenes are meant to represent King Narmer, who receives a tribute in some ceremonial context.

The tribute presented to King Narmer consists of bulls, goats, and captive prisoners. The numbers of each of them are indicated (by using the same symbols as later in hieroglyphic inscriptions) as follows: 400.000 ( ) bulls, 1.422.000 ( ) goats, and 120.000 ( ) captives. It is obvious that the numbers used in the tribute are too large to represent actual numbers of a real tribute that could have been presented to King Narmer. The high numbers stated on the mace were meant to impress by their size and to convey the power attributed to King Narmer.

A similar use of numbers may be shown on his palette with its scenes that relate to the unification of the Two Lands.[12] Among the depictions is one of King Narmer smiting a captive

---

9   Baines, "Writing: Invention and early development," p. 883, and Bard, "Origins of Egyptian writing," p. 302.
10   Schott, *Ursprung der Schrift*, pp. 1729–30.
11   Bard, "Origins of Egyptian writing," p. 303.
12   For an image of the complete palette (and an interpretation of the scenes depicted on it and similar scenes), see Davis, *Masking the Blow*, pp. 162–63.

FIGURE 4: Detail from the Narmer palette (drawing by Nadine Eikelschulte)

foreigner. Above this scene, the god Horus is shown holding the head of a foreigner on a rope. Between Horus and this foreigner a bundle of six lotus flowers is placed—possibly indicating the number of foreigners smitten by the king as 6000, again, a rather impressive achievement (figure 4).

Another early object with written numbers is the inscribed flint UC 27388 kept at the Petrie Museum in London (figure 5). The flint originates from a funerary context, this time from the area of Giza. In 1904, a tomb from the time of King Djet of the First Dynasty (ca. 3000–2890 BCE)—a seal impression with the name of this king was found in the tomb— was found and excavated by Barsanti and Daressy.[13] The mastaba and its fifty-six surrounding subsidiary burials were located approximately 1.5 miles south/southeast of the Great Pyramid; however, their exact location is unknown today. The object UC 27388 was found in one of the subsidiary tombs, together with two palettes and some limestone flakes. Palettes were at all times the tools of the trade of scribes. Early examples, like the ones found in this subsidiary tomb, are simply small tablets with two hollows to take the red and black ink. The hieroglyph for scribe (𓏞), showing a scribe's kit consisting of a palette, a bag for powdered pigments, and a holder for reeds, is attested from the late First Dynasty.[14] The limestone flakes—possibly writing material—did not show any writing. The flint, however (although not usually used to take inscriptions), shows several numbers written in the hieroglyphic style of later times. It is not known why the scribe used this flint as writing material, nor do we know what he recorded

13   Excavation report: Daressy, "Edifice"; see also the publication Petrie, *Gizeh and Rifeh*, pp. 1–6 and pl. VI.
14   See Kahl, *Hieroglyphenschrift*, pp. 833–35.

FIGURE 5: Inscribed flint UC 27388 (Courtesy of the Petrie Museum of Egyptian Archaeology, UCL)

with this set of numbers. But with this object, we have evidence that from earliest times on, numbers were part of the concerns of a scribe.[15]

The power of written numbers to indicate the existence of specific amounts of things is also used in later objects on a regular basis: the offering lists. The dead were supposed to need regular provisions of food and drink, which were guaranteed by putting depictions of these items and their quantifications in stone within the tomb, as in the example from the tomb of Princess Nefertiabet from the Old Kingdom (figure 6).[16]

Princess Nefertiabet, probably a daughter of Cheops (Fourth Dynasty), is depicted in front of an offering table with loaves of bread. Around the table, further offerings are depicted, and the sign for 1000 (𓆼) is written three times—presumably to indicate respective quantities. Likewise, the list of fabrics to the right of the scene indicates types of cloth and their respective dimensions and amounts.

Early evidence for the use of number notation originates mostly from funerary or temple contexts. However, this may mostly be due to the vagaries of preservation. The usage of numbers in a funerary context can be described as administrative and may serve as a mirror for the usage of numbers in daily life at this time. Apart from the administration of goods, in

---

15 Another scribe also took tools of his trade to his tomb—the oldest extant papyrus (uninscribed) originates from Saqqara from the tomb of Hemaka, a high official under king Den of the First Dynasty, See Emery, *Hemaka*, and Wilkinson, *Early dynastic Egypt*, p. 147.
16 For a detailed description of the object cf. Ziegler, "La princesse Néfertiabet."

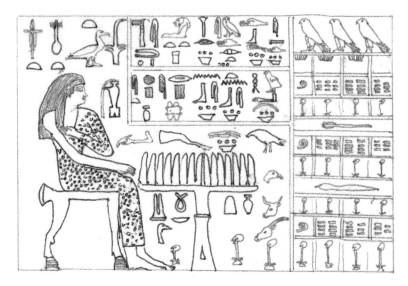

FIGURE 6: Stela of Princess Nefertiabet (Old Kingdom) Louvre E 15591 (drawing by Nadine Eikelschulte)

which numbers and quantities have an inherent place, numbers also obtain a secondary, ritual meaning, which is shown by objects like the Narmer macehead and Narmer palette but also by the offering lists. The writing of numbers is used to indicate the presence of something and to specify its quantities. With the two royal objects, extremely large numbers are used to indicate overwhelming power. The offering lists secure the existence of specific (and not small, either) numbers of goods for the deceased.

4.

Summary

When writing was developed in Egypt at the end of the fourth millennium BCE, it was at least partly due to administrative needs. Therefore, it is not surprising to find representations of numbers among the earliest evidence of written material we have from Egypt. The evidence from the tomb U-j reflects the administrative function of script and numbers, which by then had been taken over into the funerary context (in form of the "administration," i.e., the recording of grave goods).

Even before the First Dynasty, the number system was fully developed. It used a set of seven distinct hieroglyphic signs to represent powers of 10 in a decimal, nonpositional number system. Even the earliest evidence we find in tomb U-j already shows two of these seven signs (the hieroglyphs to denote 1 and 100). In the same way that writing is monopolized to serve the elite (i.e., ultimately the king), numbers are appropriated. Apart from their utilitarian use in administration, they are also employed in representational contexts, as is demonstrated by the macehead (and perhaps also the palette) of King Narmer. Earliest evidence of the burial of a scribe identified by the tools of his trade among his grave goods includes an object that demonstrates explicitly his numeracy.

The evidence presented in this chapter points to the close connection between literacy and numeracy, which will resurface again at later times in the history of Egyptian mathematics.

OLD KINGDOM

The Old Kingdom is generally considered the first cultural peak in the history of ancient Egypt.[1] The output in art, architecture, and literature all bear witness to the economic prosperity and political stability of this period. During the first part of the Old Kingdom (Dynasties 3 and 4), Egypt enjoyed almost complete self-sufficiency without serious threats from abroad. Military campaigns served to bring in additional resources. The country was run centrally from its capital at Memphis, at the border of Upper and Lower Egypt. Local rulers of limited authority controlled the provincial administration in the individual nomes. The king was the theoretical owner of all its resources, and, moreover, he was the mediator between gods and the people. This role, which did not end with the king's death, prompted the development and execution of large-scale stone architecture, beginning with the step pyramid of King Djoser (2667–2648 BCE) in the Third Dynasty. The architect of the step pyramid, Imhotep, was later deified as the patron of scribes and physicians, indicating the recognition of the importance of writing. Writing at all times included the capability to handle numerical values as well; and obviously this capability was needed to plan and run these large-scale building projects.

Literacy and numeracy were essential prerequisites for a bureaucratic career, as is also expressed in the scribal statues, which were introduced toward the end of the Fourth Dynasty: functionaries chose to be depicted reading from or writing on a roll of papyrus.[2] This type of statue, which was predominantly popular among high-ranking officials, documents the importance of scribes in the achievements of the Egyptian state and, at the same time, their own awareness of their importance. The most famous of these statues, today known by its present

---

1   For a well-illustrated introduction and overview of this period, see Malek/Forman, *Shadow of the Pyramids*; for a recent summary, see Malek, "Old Kingdom."
2   Malek/Forman, *Shadow of the Pyramids*, p. 9.

location as the "scribe of the Louvre,"[3] belongs to the best private sculpture made during the Old Kingdom.[4]

Despite the existence of written texts from the Old Kingdom (2686–2160 BCE) on, the rate of literacy in ancient Egypt is difficult to assess. It certainly varied from period to period, probably even more so depending on the respective region (estimates vary from less than 1% to 15%). A general estimate, difficult to achieve at all because of the patchy evidence, is not likely to represent the actual situation for a specific place like Middle Kingdom Lahun (for which the 15% may be more accurate) or New Kingdom Deir el Medina—the two places from which the majority of extant written evidence on papyrus or ostraca originated. These estimates of literacy in ancient Egypt indicate that the ability to read and write was something that was treasured by a small group, to whom the option was open to become part of the royal or temple administration. They wrote the handbooks on mathematics, medicine, and other subjects that are our main source of information for scholarly knowledge in ancient Egypt. No such handbooks originating from the Old Kingdom have survived. However, a variety of sources, such as depictions in tombs, archaeological finds, and two extant archives, demonstrate the development of metrological systems; and, with the remains of the Giza pyramids still in place, it seems unlikely that mathematical techniques to plan a building project were not in place by that time.

Unfortunately, the surviving written evidence is extremely scarce. No mathematical texts are extant, and we do not have any documentation of a building project except for construction marks on the actual site. Only two archives, both of temples, have survived. These show the use of a tabular layout to administer data, and they demonstrate the knowledge of various metrological systems. However, these are too fragmentary to derive any concrete mathematical techniques. Further evidence that mathematics was integral to the profession of scribe (during the Old Kingdom and later) can be found in the autobiographies of the scribes of the Old Kingdom. Autobiographies of officials in their tombs are attested throughout pharaonic history; they serve to secure the afterlife of the deceased official by ascertaining in writing that the life of the deceased had proceeded according to *Maat*, the Egyptian world order. Thus, the Egyptian autobiographies usually give an account of the professional life of the official and indicate repeatedly that the life and work of the official had happened according to the *Maat*. The autobiographic texts from the Old Kingdom emphasize regularly that the king expressed satisfaction with the work of a scribe and that they were rewarded and promoted by the king

---

3   Louvre E3023, probably Fourth Dynasty.
4   Malek/Forman, *Shadow of the Pyramids*, p. 8.

for the successful completion of their tasks, which are mentioned or briefly described and which can often be linked to the use of numeracy, for example, when a scribe is praised for his administrative achievements in controlling the king's revenue.

The invention of a calendar is another accomplishment for which ancient Egypt is to be credited by the time of the Old Kingdom.[5] A lunar calendar, which is believed to be the first calendar established in Egypt, was based on the observation and, supposedly, the record of celestial phenomena, that is, the heliacal rising of Sirius (or Sothis) to mark the beginning of the year and of the moon to mark individual months. The rising of Sirius is supposed to have been a significant event (more so than the helical rising of other stars during the year) because it was linked with the most important yearly event in agriculture, the annual flooding of the Nile. The Egyptian state depended on the administration and control of agricultural products by the state and temples. The agricultural year comprised three seasons, beginning with the heliacal rising of Sothis (around July 19 in our calendar),[6] which indicated the imminent Nile flood and, hence the beginning of the season of inundation (*3ḥ.t*). Once the water had receded enough to permit the sowing of seed, the second season began, in which seeding, tilling, growth, and harvest took place. This was followed by a period of low water, the dry season of the summer. Each of these seasons lasted approximately 4 months. The Egyptian agriculture relied on one harvest per year, and predictions about this harvest could be made according to the level of inundation, which was recorded from early on. In order to control these agricultural cycles and to carry out a regular census, some instrument of measuring the individual periods, that is, a calendar, was needed. The lunar calendar can be traced only via feast days, which were determined on the basis of a lunar cycle as well as via the names of the months. It supposedly consisted of 12 months of either 29 or 30 days, with an extra intercalary month every 3 years or so to bring the calendar into accord with the actual seasons. It is assumed that the supposed drawback of this calendar (for administrational practice), the irregularity caused by the leap months to be inserted (according to observation?), led to the development of the other Egyptian calendar, the so-called civil calendar, used in administration. This calendar consisted of three "seasons," each comprising 4 months of 30 days, and, at the end of the year, 5 extra days (called "epagomenal") were added. This calendar was probably introduced between approximately 2937 and 2821 BCE,[7] and then the beginning of the year coincided with the

5   For comprehensive accounts of calendars in ancient Egypt, see Parker, *Calendars*, Spalinger, *Revolutions in Time*, and Depuydt, *Civil Calendar*.

6   The heliacal rising of Sirius (or Sothis) is indicated by the first visibility of this star at dawn after its period of invisibility. For a detailed explanation, see Parker, *Calendars*, p. 7, sec. 21.

7   Parker, *Calendars*, p. 53.

heliacal rising of Sirius indicating the yearly inundation. As this civil calendar is about $\frac{1}{4}$ day shorter than the solar year, however, every 4 years it moved by one day against the solar year. After 1460 years the civil calendar and the solar year were in accordance again. While this movement of the calendar against the natural year may have been awkward at times, there can be little doubt that the practical advantages of 12 months all of the same lengths were considerable for its administrative function. Thus, the civil calendar remained unchanged throughout Egyptian history, and the attempt to introduce a sixth epagomenal day every fourth year in 237 BCE did not have any effect. At the same time, a lunar calendar continued to be used in temples—the other main area where timekeeping would be required in order to perform the necessary rituals and festivities at their appropriate times.

5.

The Cultural Context of Egyptian Mathematics in the Old Kingdom

Although there are no mathematical texts extant from the Old Kingdom to inform us in detail about the mathematical techniques available at that time, a number of sources provide information about the kind of mathematics and its context at that time. At least indirect evidence for the use of mathematics in administration can be drawn from the Abusir papyri, which originate from the mortuary temples of two kings of the Fifth Dynasty at Abusir.[1] They document the running of a mortuary temple and include duty rosters for priests, lists of offerings and inventories of temple equipment, and letters and permits. These texts also indicate the assessment of cattle at regular intervals. With regard to formal layouts, several examples of the Abusir papyri, such as papyrus UC 32366f, which was probably a table of duty for staff on rotating service, show the arrangement of numerical and other data in tabular format. The use of tabular formatting is fully developed in the Old Kingdom.[2] The table of papyrus UC 32366f is a formal table, each horizontal line corresponded to one day and each group of vertical columns, to a particular duty in the temple. The cells could hold the name of the individual whose task occurred on a specific day or the information that a task had not been completed, which was indicated by a red stroke. The mathematical regularity expressed in the tabular format of this example was presumably that of consecutive days running through the rows of

---

1   The papyrus archive of Abusir was published in Posener-Kriéger/de Cenival, *Hieratic Papyri* (photos and hieroglyphic transcriptions) and Posener-Kriéger, *Archives du temple* (translations and commentary). For a comprehensive account of the history of Abusir, see Verner, *Forgotten Pharaohs*, or Verner, *Abusir*.

2   The terminology used in the following description of tables is that of Robson, "Tables," p. 20: "I shall define a formal table as having both vertical and horizontal rulings to separate categories of information; informal tables, on the other hand, separate quantitative and qualitative data by spatial arrangement only, without explicit delimiters. Documents with no tabular formatting at all we might call prose-like or prosaic. Headed tables have columnar headings, while unheaded tables do not."

the table. Supposedly, although this can no longer be ascertained due to the fragmentary state of the papyrus, this was also a headed table, with the tasks written in red in the header.

Even older than the Abusir papyri are the Gebelein papyri, a set of administrative documents dating to the Fourth Dynasty and so far the oldest written papyri that have been found. They were discovered in a wooden box, together with writing equipment in an anonymous Old Kingdom tomb at Gebelein.[3] These papyri contain information about the social and economic life in Gebelein during the Fourth and at the beginning of the Fifth Dynasties. Various documents from these papyri are written as tables, including, for example, an account of four grain products that are allocated to ninety-one persons, which covers the complete front of Papyrus Gebelein II.

How mathematical techniques to administer quantities of grain, to plan building projects, and others developed or what they were exactly at this time, we cannot say. Nevertheless, some scribes of the Old Kingdom (whom we assume to have developed some mathematical knowledge) have left descriptions of their lives and careers within their tombs that at least allow assessing the cultural environment in which they worked.[4] There is evidence that the sons of scribes followed their fathers in their careers; for example, the inscription of Akhethotep, an official of the Fifth Dynasty, mentions a tribute by his son "for having educated him to the satisfaction of the king";[5] the description of the career of Harkhuf begins with an expedition to Yam, where he is sent with his father, followed by a second expedition to Yam, where he is then sent on his own.

The autobiographic texts of officials from the Old Kingdom emphasize regularly that the king expressed satisfaction with their work and that they were rewarded and promoted by the king for the successful completion of their tasks. The following excerpt taken from the autobiography of Weni from the Sixth Dynasty (2345–2181 BCE) may serve as an example. Weni was an official whose career spanned the reigns of three kings (Teti, Pepy I, and Merenra). His autobiography is written down on a monolithic slab of limestone, which formed one wall of his single room tomb chapel.[6] Weni praises himself for having done the following:

---

3   See Parkinson, *Papyrus*, p. 47. The papyri were published in Posener-Kriéger and Demichelis, *Papiri di Gebelein*. See also Posener-Kriéger, "Papyrus de Gébelein."
4   For a list of ninety-eight autobiographies from the Old Kingdom and their respective publications, see Kloth, *Autobiographische Inschriften*, pp. 3–44.
5   See Strudwick, *Pyramid age*, p. 261 (no. 194).
6   The inscription is published in Mariette, *Abydos*, plates 44–45. The hieroglyphic text is edited in Sethe, *Urkunden des Alten Reichs*, pp. 98–110; but see also the new edition in Hofmann, "Autobiographie des Uni", pp. 226–28. An English translation can be found in Lichtheim, *Literature I*, pp. 18–23 and Simpson, *Literature*, pp. 402–7. For a discussion of Weni's career in the context of the social and political structures of the Old Kingdom, see Eyre, "Weni's career."

When I was chamberlain of the palace and sandal-bearer, King Mernere, my lord who lives forever, made me Count and Governor of Upper Egypt, from Yebu in the south to Medenyt in the north, because I was worthy in his majesty's heart, because I was rooted in his majesty's heart, because his majesty's heart was filled with me. When I was chamberlain and sandal-bearer, his majesty praised me for the watch and guard duty which I did at court, more than any official of his, more than any noble of his, more than any servant of his. Never before had this office been held by any servant. I governed Upper Egypt for him in peace, so that no one attacked his fellow. I did every task. I counted everything that is countable for the residence in this Upper Egypt two times, and every service that is countable for the residence in this Upper Egypt two times. I did a perfect job in this Upper Egypt. Never before had the like been done in this Upper Egypt. I acted throughout so that his majesty praised me for it.[7]

From autobiographic texts like this, we grasp that the tasks of an expert in the royal service were many and varied. While we expect Weni to have had mathematical knowledge to enable him to control and record the results of the assessment of products and services that he claims to have executed twice, he must also have been able to keep up a certain order in the regions he was supposed to control. Other parts of his work included some guard service at court; the following sentence seems to imply that he also had a similar function in the region that he was given to administer. Throughout the text, Weni makes reference to the satisfaction of the king with his work, and it is by the king that he is appointed to his functions. The indication that he executed the counting of services and produce twice ("countable" refers to those items that result in revenue for the king) suggests that this may have happened at regular intervals.[8]

The autobiographic inscription of Ankh-Meryre-Meryptah includes several references to mathematics and its context(s):

(1) [The Sole Companion, Royal Architect, Ankh-Meryre-Meryptah] he says:

(2) [I am a builder for King] Meryre, my lord. His majesty sent me to direct the work of his monument in On. I acted to the satisfaction of his majesty. I spent

7   Translation from Lichtheim, *Literature I*, p. 15.
8   The fact that Weni claims to have done it twice has also been interpreted as a description of Weni's treatment of his inferiors: Gardiner assumed that Weni "squeezed out of the unfortunate inhabitants of Upper Egypt twice as much in the way of taxes and work as his predecessors"; see Gardiner, "Regnal years," p. 15.

six years there in directing the work, and his majesty rewarded me for it as often as I came to the residence. It all came about through me by the vigilance I exercised

(3) - there through my own knowledge. . . .

When I was in the service of my brother, the Overseer of Works -, I wrote and I carried his palette. Then, when he was appointed Inspector of Builders, I carried his measuring rod.

(5) Then, when he was appointed Overseer of Builders, I was his companion. Then, when he was appointed Royal Architect-Builder, I governed the village for him and did everything for him efficiently. Then, when he was appointed Sole Companion and Royal Architect-Builder in the Two Administrations, I reckoned for him all his possessions, and the property in his house became greater than that of any noble's house.

Then, (6) when he was appointed Overseer of Works, I represented him in all his affairs to his satisfaction with it. I also reckoned for him the produce of his estate (pr-ḏt) over a period of twenty years. Never did I beat a man there, so that he fell by my hand. Never did I enslave any people there. As for my people there (7) with whom I had arguments, it was I who pacified them. It was I who gave clothing, bread, and beer to all the naked and hungry among them. . . . [9]

The work of Ankh-Meryre-Meryptah as a builder included—as stated in the beginning of the quote—that he directed building projects; hence it may be fair to assume that he had to determine amounts of building materials as well as plan the work of the team. He explicitly refers to the use of his knowledge, which presumably included mathematical knowledge of some kind. We also learn that he was trained by a member of the family; in several autobiographies, the father is mentioned as the teacher, in the case of Ankh-Meryre-Meryptah, his elder brother was the one who trained him. The brother was also working on building projects, and Ankh-Meryre-Meryptah states explicitly that as an apprentice of his elder brother, it was his job to carry the measuring rod—again a reference to mathematical occupations, assuming that he not only carried the measuring rod but used it and worked with the results obtained from taking measurements. In addition to assisting the brother in his work, Ankh-Meryre-Meryptah also stated that he did the bookkeeping for his brother, another occupation that implicitly necessitates the use of mathematical techniques, and according to the statements in his autobiography, Ankh-Meryre-Meryptah did this exceedingly well.

---

9   Translation from Lichtheim, *Autobiographies*, pp. 12–13. The text is also included in Strudwick, *Pyramid Age*, p. 265–67 (No. 198).

The succession in office is also the framework of one of the teachings, often attributed to the Old Kingdom (possibly the latter part of the Sixth Dynasty), although the surviving manuscripts date to the Middle and New Kingdoms only. The *Teaching of Ptahhotep* begins with the statement of the mayor of the city, the vizier Ptahhotep, who asks the permission of the king to train a successor ("a staff of old age"), because he himself has become old:

May this servant be ordered to make a staff of old age, so as to tell him the words of those who heard, the ways of the ancestors, who have listened to the gods.[10]

The teaching states *expressis verbis*, at the beginning as well as at the end, the necessity of the king's approval to appoint a successor. The maxims taught in the *Teaching of Ptahhotep* include aspects of professional life (i.e., the correct behavior toward superiors and inferiors within the administrational system, the advice to accept one's position in life, and a warning against exploiting power or acting greedily—again this would be something a scribe would be able to do because of his numerical proficiency) as well as private life (e.g., the advice to take a wife and treat her well).

The option to become a scribe was obviously limited to those persons who were able to read and write (which included the ability to handle numerical values), which presumably was taught by fathers to their sons (or, as in the example of Nekhebu, within a family from brother to brother). This prerequisite is also expressed in the basic Egyptian designation of these experts, "scribe." Hence, a certain likelihood of becoming a scribe was guaranteed by being a member of a scribal family. This very basic prerequisite is then also expressed in a form of self-depiction chosen by high officials from the Old Kingdom on: the scribal statue. Scribal statues depicted an official in the process of reading or writing a document. Other depictions in tombs of high officials often include scenes of the tomb owner inspecting the harvest. This, also, would presuppose the skillful handling of numbers, that is, quantities of goods measured in a variety of metrological systems. At the same time, lower ranks of scribes are depicted preparing the respective documentation, and, judging by depictions of scribes being beaten, the documentation apparently was not always considered to be appropriate.

The use of mathematics by the scribes or experts from the Old Kingdom can also be deduced from the variety of official titles attested in this period.[11] Many point to mathematics as part of the daily work of their holders. This includes the administration of offerings and

---

10   Lichtheim, *Literature I*, p. 63.
11   For a collection of titles of the Old Kingdom and their attestations see Jones, *Titles*. See also Baer, *Rank and Title*, and Strudwick, *Administration*.

FIGURE 7: Supervision of baking and brewing from the tomb of Ti (Fifth Dynasty). (Drawn after Wengrow, *Early Egypt*, page 96)

produce of various kinds, as well as the bookkeeping of royal storehouses. Noteworthy in this respect are the various scribes, for example, *zš ꜣḥwt*, "field scribe,"[12] *zš ꜣpdw*, "scribe of the aviary,"[13] and *zš šnwt pr-ḥḏ*, "scribe of the granary (and) of the treasury."[14] Several of these titles hint that the respective "writing" was mostly concerned with accounting and, therefore, had to involve mathematics. An explicit reference to accounting (*ḥsb*) is found in the title of the *zš ḥsb qd-ḥtp* "accounts scribe of the builders of the offering places."[15]

In addition, accounting and counting by scribes is depicted in various tomb decorations—for example, the supervision of baking and brewing (figure 7) or the inspection of cattle and birds from the tomb of Ti at Saqqara. Ti was an official who lived during the Fifth Dynasty (2494–2345 BCE) and worked in the immediate vicinity of the king. His income from his positions and his private possessions enabled him to construct a large mastaba for himself and his family, his wife Neferhetep, and their son Demez. All its cult rooms were decorated in painted relief depicting various aspects of the life of a high official.[16]

From this evidence, it seems fair to conclude that mathematics constituted an essential tool in the daily works of a scribe. Its use seems to be most obvious in a project like the construction of the Giza pyramids. Likewise, the importance of metrology within the mathematical abilities of a scribe becomes obvious. An important part of the daily use of mathematical techniques by the scribes in administering all kinds of goods must have included the handling of the respective metrological systems attached to these goods.

---

12   Jones, *Titles*, no. 3043.
13   Ibid., no. 3042
14   Ibid., no. 3204.
15   Ibid., no. 3159
16   For a publication of this tomb, see either Steindorff, *Grab des Ti*, or Epron, Daumas, and Goyon, *Ti*, and Wild, *Ti* (French publication).

6.

Metrological Systems

The development of metrology constituted the foundation of the quantitative control of agricultural resources, which then enabled the cultural achievements of ancient Egypt.[1] Note that while some units remained the same and were used throughout Egyptian history, others became obsolete or changed. This chapter focuses on metrological units, which are attested in the Old Kingdom.[2] Some metrological units were linked to a specific kind of object to be measured; thus it is to be expected that there were different units for assessing volumes of grain, liquids, or building materials.

6.1 LENGTH UNITS

Egyptian units of length were generally derived (as in many other cultures) from parts of the human body. The basic measuring unit was the cubit (*mḥ*), which was derived from the length of a person's forearm, that is, the part of the arm from the elbow to the fingers.[3] Based on extant cubit rods, the Egyptian cubit is indicated in modern literature to equal approximately 52.5 cm. However, it is not to be assumed that a standard cubit of exact length existed; local variations have to be taken into account. The cubit rods used to establish the average length are those of the New Kingdom. Hence this average length is to be taken as a rough guideline,

---

1   For a similar role of metrology in Mesopotamia, see Robson, "Literacy."
2   Only few comprehensive studies of metrological units exist so far; for example, Roik, *Längenmaßsystem*, for length measures, Pommerening, *Hohlmaße*, for capacity measures, and Cour-Marty, "Poids égyptiens," for weights. See also Reineke, *Gedanken und Materialien*, pp. 71–130 and pp. 144–93.
3   For an overview of length measurements, see also Pommerening, *Hohlmaße*, pp. 271–74.

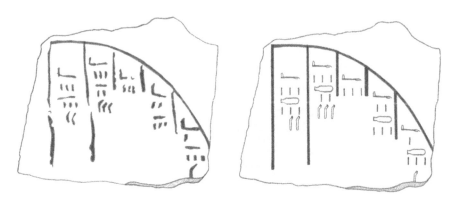

FIGURE 8: Ostracon (Cairo JE50036) with geometrical sketch

not an exact standard. The cubit was divided into 7 palms (*šzp*), and each palm was made up of 4 fingers (*ḏbꜥ*).[4]

An ostracon (figure 8), presumably from the Third Dynasty, displays the use of the cubit system.[5] This ostracon was found at Saqqara near the step pyramid of King Djoser. It probably depicts a vaulted roof, indicating its height at several (presumably equidistant) points. According to the measurements noted on the drawing, the highest point of this vault measured 3 *mḥ*, 3 *šzp*, 2 *ḏbꜥ* (i.e., approximately 183.8 cm); its lowest point was 1 *mḥ*, 3 *šzp*, 1 *ḏbꜥ* (i.e., approximately 76.9 cm). Transferring this drawing into a modern technical sketch illuminates the problems that this kind of source presents. If we try to establish a drawing to scale out of the data noted on the ostraca, we can easily sketch the five vertical lines, as their individual lengths are explicitly marked in the Egyptian drawing as 3 cubits 3 palms 2 fingers, 3 cubits 2 palms 3 fingers, 3 cubits, 2 cubits 3 palms, and 1 cubit 3 palms 1 finger.

However, the drawing does not indicate at which distance these measurements were taken. If we assume that they were equidistant and if we further assume that they were spaced 1 cubit apart, a drawing like that on the right side of figure 8 results, which happens to look similar in shape to the Egyptian drawing on the ostracon. As Corinna Rossi states:

---

4   For evidence of a further unit of length used in the Old Kingdom, the *ṯbt*, attested only twice and only in the mastaba of Ptahshepses, see Žába, *Ptahshepses*, p. 83 (graffito no. 18) and Verner, "Unbekanntes Maß."

5   For the construction of this figure and a discussion of related objects, see Rossi, *Architecture and Mathematics*, p. 115–16. Even earlier evidence for the use of length measurements can be found on jars from the Second Dynasty, see Ritter, "Metrology," p. 29.

If the shape of the object to be represented was clear, an eyeball sketch could reproduce its proportions with a remarkable degree of accuracy, without necessarily being a conscious, calculated scale drawing.[6]

However, numerous examples throughout Egyptian history demonstrate that Egyptian drawings are not to scale.[7] Rather, their annotations have to be read carefully in order to interpret the drawing. And, as in the case of this drawing, there were not enough annotations for us to create a drawing to scale; hence, some caution is needed in interpreting the original. Guesses about scale aside, the drawing still allows a number of observations. First of all, it demonstrates the ability to mathematize a curved line by measuring its height at various points. Obviously, the measurements were all taken from a baseline (assuming, yet again, that the lengths indicated are referring to the lines to the left of them). Finally, it could be argued in this case that the conditions we assumed in producing our modern sketch to scale, that is, the vertical lines being equidistant and spaced 1 cubit apart, were conventions and therefore would not have to be noted explicitly.

Apart from architecture, length measurements were also of use for administrative purposes, for example, to record the height of the Nile flood in order to set the taxes for the year.[8] Early evidence of this can be found on the so-called Palermo stone (figure 9).[9] The Palermo stone is the modern designation (named after the place where it has been held since 1877) of the largest piece of a fragmented stela, which dates to the end of the Fifth Dynasty.[10] The Palermo stone measures 43.5 cm in height and 25 cm in width, and its thickness varies between 5.1 and 6.5 cm.[11] The complete stela is estimated to have measured 2.1 m in height and 60 cm in width, with a thickness of 6.5 cm. Only parts of it are extant: apart from the Palermo stone, there are further fragments at the Egyptian Museum in Cairo and at the Petrie Museum in London. The origin of the stone is unknown.[12] The black stone (basalt?) is inscribed on both

---

6   Rossi, *Architecture and Mathematics*, p. 112.
7   For a discussion with several examples, see Rossi *Architecture and Mathematics*, pp. 101–13.
8   For a discussion of measuring the Nile flood, see Seidlmayer, *Nilstände*.
9   The most recent translation is Strudwick, *Pyramid Age*. The description and translation of the Palermo Stone and its related fragments given in Wilkinson, *Palermo Stone* should be read with care; see the review in Baud, "Review Wilkinson."
10   It is discussed, however, that the extant object may be a copy from the time of the Twenty-fifth Dynasty (747–656 BCE), see Helck, "Palermostein."
11   Wilkinson, *Palermo Stone*, p. 18.
12   Porter and Moss: *Topological Bibliography IV*, p. 133, lists "the neighborhood of el-Minya according to native report" as provenance of three fragments now in Cairo; another one of the Cairo fragments was supposedly found in Memphis. However for the (un)reliability of this kind of information, see Godron, "Pierre de Palerme," p. 19, quoting Petrie: "While I was at Memphis, the sixth piece was offered to me by a petty dealer. I at once bought it, and then handed it to a confidential native to show about quietly, and make inquiries. The story, as gradually obtained, was that it was found in Upper Egypt, been brought

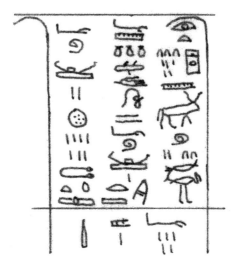

FIGURE 9: Section of the sixth register of the Palermo stone (drawing by Nadine Eikelschulte)

sides. The extant fragments contain records of prehistoric (possibly fictitious) kings and year names from the reigns of Aha, the first king of the First Dynasty (ca. 3000–2686 BCE) until at least Neferirkare, the third king of the fifth Dynasty dynasty (2494–2345 BCE). The annals are arranged in registers of rectangles. Each year is recorded by a single event. From King Djer (second king of the First Dynasty) onward, the height of the Nile is recorded for each year in the lower part. The measurements are given in the cubit system up to fractions of a digit. Note that the side borders of the section are a stylized version of the hieroglyph *rnp.t* "year" ( ). Consequently, the left border should be read with the next entry. The entry for the year on figure 9 reads

> Setting up 35 estates with people and 122 cattle farms; construction of one ship of cedar/pine wood (of type or name) 'Adoring the Two Lands,' of 100 cubits and two ships of 100 cubits of *meru*-wood; seventh occasion of the count.[13]

The height of the Nile is indicated as 5 *mḥ*, 1 *šzp*, 1 *ḏbꜥ* (5 cubits, 1 palm, 1 finger, that is, approximately 2.72 m). At the end of the main entry, a "count" is mentioned—that is, yet

---

down and sold to a Cairo dealer: from him it had been passed to a Memphite dealer, and so finally to myself. This shows how one piece at Memphis had come from elsewhere, and the Cairo-Memphite piece may have a similar history."
13   Strudwick, *Pyramid Age*, p. 66.

FIGURE 10: Scene from the tomb of Nianch-Chnum and Chnumhotep

another numerical aspect of this early period. From the Palermo stone it seems that within the reign of a king, a recurring census of cattle was carried out; at exactly which intervals or what kind of event led to individual counts is not fully understood. It seems to have taken place approximately every 2 years.[14] The data collected (one would assume the number of adult and young cattle) and their possible implications (a tax in cattle to be given at the end of a certain period?) are not clear.

Further evidence for the use of the cubit can be found within representations of trade and scenes depicting administration of grain in the decorations of tombs of high-ranking officials—which will also be the main source for area and capacity measures. In the process of harvesting grain, it is sometimes shown to be piled into a heap, as in the scene from the tomb of Nianch-Chnum and Chnumhotep (figure 10). The accompanying inscription reads: "Piling up a heap of 60 cubits."

Another scene from the same tomb (figure 11) shows the sale of a piece of cloth, measured in cubits, with the comment: "1 + x cubits of cloth for payment of six *shat*. I tell you this truly—it's a *netjeru* cloth, really careful work."[15] The evidence presented here documents a wide range of uses for the cubit system in various contexts: architecture, administration, and trade. The cubit was also the basis on which a system of metrological units to measure areas was founded.

---

14   For a discussion, see Strudwick, *Pyramid Age*, p. 11.
15   Translation of Strudwick, *Pyramid Age*, p. 410.

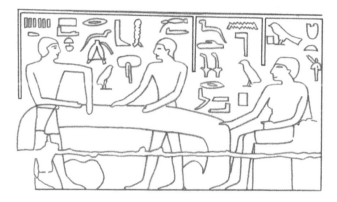

FIGURE 11: Scene from the tomb of Nianch-Chnum and Chnumhotep

6.2 AREA UNITS

Area measures are related to the basic length unit, the cubit. The smallest unit is again called "cubit," *mḥ*: it designates a square of 1 *mḥ* by 1 *mḥ*. The next bigger unit is the *t3* ("land unit"), which comprises a square of 10 *mḥ* by 10 *mḥ* (about 27.65 m²), followed by the *ḫ3*, a strip of 10 *mḥ* by 100 *mḥ* (about 275.65 m²) and the *sꜣ.t*, a square of 100 *mḥ* by 100 *mḥ* (about 2756.5 m²). The *t3* is further divided by a set of submultiples called *rmn* ("shoulder," equal ½ *t3*), *ḥsb* ("account unit," equal 1/4 *t3*) and *z3* (equal 1/8 *t3*).[16] For an overview of the respective signs see table 2.

Note that there are considerable differences between area measures in the Old, Middle, and New Kingdoms.[17] Area measures are used in the Old Kingdom to measure the size of fields, as in the example from the tomb of Metjen.[18] Metjen was a nomarch (the highest official of an Egyptian administrative district called a nome) who began his career during the reign of King Huni (2637–2613 BCE), the last king of the Third Dynasty; he died under King Sneferu (2613–2589 BCE), the first king of the Fourth Dynasty. His tomb in Saqqara is decorated with inscriptions in form of titles and official documents, which contain detailed information

---

16 A list of references for the individual units can be found in Baer, "Egyptian units of area, pp. 113–14.
17 The basic unit during the Middle Kingdom is called *mḥ-t3* and designates a rectangle of 1 *mḥ* by 100 *mḥ*. This is followed by the *sꜣt* = 100 *mḥ-t3*, a square of 100 *mḥ* by 100 *mḥ*. The largest unit is called *ḫ3-t3* = 10 *sꜣt* = 1000 *mḥ-t3* = 1000 *mḥ* by 100 *mḥ*. See Gödecken, *Meten*, p. 360.
18 Inscriptions of this tomb have been published (in German) in Gödecken, *Meten*.

TABLE 2: Attested area measures from the Old Kingdom (after Baer, "Egyptian units of area," p. 115

Egyptian	⌐◡	𓎝	✶	◡―	▬	𓈖	▬
transliteration	*mḥ*	*z3*	*ḥsb*	*rmn*	*t3*	*ḫ3*	*sꜣ.t*
translation	cubit		account unit	shoulder	land unit	thousand	aroura
size (cubit × cubit)	1 × 1				10 × 10	10 × 100	100 × 100
size (in relation)		$\frac{1}{8} t3$	$\frac{1}{4} t3$	$\frac{1}{2} t3$			

about his career (especially his promotions), possessions, and income.[19] The fourth decree from his tomb reads as follows:[20]

Decree IV

(To) the overseer of commissions of the fourth/fifth and third nomes of Lower Egypt:

12 'foundations of Metjen' have been founded for him in the fourth/fifth, sixth and second nomes of Lower Egypt (along with) their products for him in the dining room. These have been bought for 20 arouras of land from many royal colonists, (along with) 100 portions of funerary offerings which come daily from the soul chapel of the royal mother Nymaathap (and with) a walled estate 200 cubits long and 200 broad, set out with fine trees, and a large pool made in it; it was planted with fig trees and vines.

It is written down in a royal document, and their names are (recorded likewise) on (this) royal document. The trees and the vines were planted in great numbers, and the vine therefrom was produced in great quantity. A garden was made for him on land of 1 *kha* and 2 *ta* within the enclosure, which was planted with trees. (It was named) Iymeres, a 'foundation of Metjen,' and Iatsobek, a 'foundation of Metjen.'

The property of his father the judge and scribe Ipuemankh was given to him, without wheat and barley and the property of the estate, but with dependents and

---

19   For an English translation, see Strudwick, *Pyramid Age*, pp. 192–94.
20   Translation by Strudwick, *Pyramid Age*, pp. 193–94.

herds of donkeys and pigs (?). He was promoted to first of the scribes of the office of provisioning and overseer of the office of provisioning; he was promoted to be the 'strong of voice' among those involved in agricultural production when the boundary official of the sixth nome of Lower Egypt was in charge of the judge and supervisors of revisionary offerings in the sixth nome of Lower Egypt, who should take the job of judge and 'strong of voice', and he was promoted to be overseer of all linen products of the king; and he was promoted to be ruler of the Per-Desu and the towns which are under the same control (?); and he was promoted to be the boundary official of the people of Buto and controller of the estate of Per-Sedjaut and Per-Sepa and boundary official of the fourth/fifth nome of Lower Egypt and controller of the estate of Senet and the nomes under the same control, controller of Per-Sheptjet, and controller of the towns of the Great Mansion of the southern lake (the Fayum).

The 'foundations of Metjen' have been founded out of what his father Inpuemankh gave him.

Apart from documenting the existence of the individual metrological units, these sources also prove that the sizes of fields (i.e., their areas) were measured during the Old Kingdom—from which one might deduce the ability to calculate the area of rectangles and possibly other geometric figures. Note that the decree indicates the size of the fields used to pay for the foundations and the size of the garden using area measures, while the size of Metjen's walled estate is indicated by its dimensions (200 cubits by 200 cubits).

6.3 CAPACITY UNITS

Concerning the measuring of volume, the extant sources are likely to omit an important part of the picture. Obviously, there is significant building activity relating to the basic length measure used in architecture, the Egyptian cubit. Unfortunately, there is no extant written evidence for the planning or carrying out of a larger construction project from the Old Kingdom.

Sources describing volume in the Old Kingdom practically all relate to quantities of grain for which a separate system (independent of the cubit) existed.[21] The standard unit for measuring cereals was the ḥq3.t, measuring approximately 4.8 l. It was divided (for smaller

---

21 For a collection and discussion of sources pertaining to grain measures throughout Pharaonic history, see Pommerening, *Hohlmaße*.

quantities) by a series of submultiples similar to that of the *t3*.[22] The submultiples are attested as early as the Second Dynasty.[23]

For larger capacities, the *h3r* was used. As with the area measures, it is important to note that relations of units change over time. While during the Old and Middle Kingdom 1 *h3r* = 10 *hq3.t*, we find in the Rhind mathematical papyrus the relation 1 *h3r* = 20 (or 5 quadruple) *hq3.t*, and from the Twentieth Dynasty on, 1 *h3r* = 16 *hq3.t*.

The fact that depictions of measuring grain are found so often in scenes in tombs of officials points to the importance of its protection and administration by the state, that is, the king. Early evidence of this control (possibly dating to King Huni) can be found on inscriptions on beer jars found on Elephantine, which supposedly once contained papyri documenting the delivery of commodities. It is assumed that the inscriptions on the vessel contained a sort of summary of the documents stored inside. The inscription of one of the vessels from Elephantine reads:[24] "Year of: The following of Horus; eleventh occasion of the count of the herds of Heliopolis. Grain, 25 *hq3.t* measures." Examples for the use of the submultiples of the *hq3.t* can be found in the Abusir papyri.

6.4 WEIGHTS

Weights were mostly used to measure precious stones and metals.[25] The basic Egyptian unit for measuring weight is the *dbn*. While the designation remains the same throughout Egyptian history, its actual weight changes significantly. During the Old Kingdom, the *dbn* weighed approximately 13.6 g.[26] Although there is practically no mention of them within texts, thousands of actual Egyptian weights are still extant. Major collections can be found at the Petrie Museum (London) and at the Oriental Institute (Chicago).

The acquisition of a weight is represented in the Old Kingdom tomb of Nianch-Chnum and Chnumhotep. The tomb contains a sequence of so-called market scenes showing the trading of various goods. In one of the scenes (see figure 12), a "worker of *dbn*" is shown handing something (a *dbn* weight?) to a person sitting on the ground and is receiving "the payment for this *dbn* of yours."

---

22   For the forms of the signs expressing the submultiples and their earlier misunderstanding as fractions of the eye of the god Horus, see Ritter, "Eye of Horus."
23   Ibid., pp. 304–5.
24   Translation of Strudwick, *Pyramid Age*, p. 74.
25   For a detailed overview see Cour-Marty, "Poids égyptiens."
26   This remains the same during the Middle Kingdom, but besides the (gold) *dbn* of 13.6 g, there appears to be also a copper *dbn* of around double its weight (27.3 g). In the New Kingdom, the *dbn* weighed 91 g and was subdivided into 10 *qdt*.

FIGURE 12: Market scene involving a weight from the tomb of Nianch-Chnum and Chnumhotep

FIGURE 13: Weight of 6 *dbn* (UC 34793)

Weights were often made from valuable materials such as serpentine or alabaster or semiprecious stones. The earliest examples date from predynastic times, indicating that the beginnings of Egyptian metrology predates the Old Kingdom. Several weights can be dated through the inscription of royal names on them. They usually show, as in the example in figure 13, the number of *dbn* units they weigh, which is 6 in this example.

The precious material used to produce some of the specimens and the inscription of a royal name on them point to the importance of carrying out a measurement. While the control of measuring instruments for the length (i.e., ropes or cubit rods) is a matter of direct comparison between two items, the process of weighing is more difficult to control because it involves not only the weight but also the scale (both of which may be tampered with). The use

FIGURE 14: Weight inscribed with royal name

of a royal name (supposedly the name of the reigning king at the time the weight is produced) may indicate the "weight of authority" that is attached to the specific item of measurement. Attested are weights with the names of Cheops (Fourth Dynasty), Userkaf (Fifth Dynasty), and Sahura (Fifth Dynasty; see figure 14).

7.

Notation of Fractions

The concept of fractions used in ancient Egypt became one of the most characteristic features of its mathematics. In the past, Egyptian fractions were treated both as clumsy and as a distinctive feature of Egyptian mathematics. Egyptian fraction reckoning was blamed as one of the reasons why Egyptian mathematics was inferior to that of its neighbor culture Mesopotamia and never advanced beyond a certain level.

There can be no doubt that fraction reckoning had to overcome some technical difficulties. This can be traced by the extant table texts, which are mostly tables for fraction reckoning. However, if one follows the Egyptian development of the concept of a fraction and attempts to understand it not in comparison with nor through the eyes of our modern system, it is possible to see Egyptian fractions not as ultimately a failure inhibiting further development, but rather as the evolution of a mathematical system into a new field. It is particularly interesting to see how tools were developed to help overcome technical difficulties. From the extant sources of later periods, we can see that fraction reckoning was considered more advanced than the manipulation of integers but that it could be done competently by advanced scribes.

The Egyptian concept of fractions, that is, parts of a whole, was fundamentally different from our modern understanding. This difference is so elementary that it has often led to a distorted analysis of Egyptian fraction reckoning, viewed solely through the eyes of modern mathematicians, who marveled at the Egyptian inability to understand fractions like we do. This is expressed in the following quotation:

We do not comprehend, that humans, when they divided 5 by 7 and thus obtained $\frac{1}{7}\frac{1}{7}\frac{1}{7}\frac{1}{7}\frac{1}{7}$, like the ancient Egyptians, did not proceed to add up these 5 similar items

TABLE 3: Basic Egyptian fractions and their notations

Value	Egyptian notation	
	hieratic	hieroglyphic
$\frac{2}{3}$		
$\frac{1}{2}$		
$\frac{1}{3}$		
$\frac{1}{4}$		

to find the much simpler expression $\frac{5}{7}$, ... We do not understand, how somebody can know that what he expresses by $\frac{2}{3}\frac{1}{15}$ forms 11 parts, which can be completed to a whole by 4 parts, but does not call it $\frac{11}{15}$.[27]

Sethe's fundamental statement was made long before the social dimension of mathematics was recognized. But although it has by now become accepted that mathematics down to its roots (i.e., numbers) has this social dimension,[28] Egyptian fraction reckoning has not been reassessed.[29]

The evolution of fractions in ancient Egypt can be traced back at least as far as the Old Kingdom. The beginnings consisted of a small group of specific fractions written by special signs (see table 3). These fractions are first attested within the context of metrological systems, that is, $\frac{3}{4}$ in $\frac{3}{4}$ of a finger and $\frac{1}{4}$ in $\frac{1}{4}$ of a *sḏt*, but they retain their notation in later times as abstract fractions.[30]

---

27 My translation of the German quote, which can be found in Sethe, *Zahlen und Zahlworte*, p. 60 with references to Erman and Eisenlohr (notes 2 and 4).
28 See, for example, Urton, *Social Life of Numbers*.
29 Some, however, have voiced doubts on the blame assigned to the use of "unit fractions"; see, for example, the following statements: " . . . the supposed impediment caused by the use of unit-fractions is largely illusory . . . , and not the result of a conceptual block" (Shute, "Mathematics", p. 351) and "The contempt that is often expressed for the system of unit fraction calculation is, I believe, not entirely justified. Unit fraction expressions can be evaluated and manipulated by a wealth of algorithms and they can convey some information, for example of magnitude and approximation, as efficiently as any other representation" (Fowler, *Mathematics of Plato's Academy*, p. 266). See also the discussions in Caveing, "Egyptian quantieme" and Høyrup, "Conceptual divergence", esp. pp. 135–36 and 142–43.
30 For the evolution of abstract numbers and fractions see Ritter, "Metrology."

From this set of earliest fractions, a general concept of fraction developed. The list of earliest fractions comprises $\frac{1}{2}$, $\frac{1}{3}$, and $\frac{1}{4}$, and it may be inferred that fractions came to be understood as the inverses of integers.[31] As a consequence, the Egyptian notation of fractions did not consist of numerator and denominator but rather of the respective integer of which the fraction was the inverse and a symbol to designate it as an inverse, that is, a fraction. The hieroglyphic sign that came to be used for this purpose, was ⟨⟩, the sign for part, which was already used in the writing of $\frac{2}{3}$ and $\frac{3}{4}$. In order to designate a general fraction (always an inverse), it was placed above the integer of which the fraction was the inverse. In hieratic, this fraction marker was abbreviated to a mere dot.

One of the first researchers of Egyptian fraction reckoning, Otto Neugebauer, devised a notational system that imitates this Egyptian notation. Fractions (i.e., inverses of integers) are rendered by the value of the integer and an overbar to mark them as fractions. Thus, $\frac{1}{5}$ would be written as $\overline{5}$, $\frac{1}{6}$ as $\overline{6}$, and so on. The only exception, the fraction $\frac{2}{3}$, was rendered by Neugebauer as $\overline{\overline{3}}$. The fractions $\frac{1}{2}$, $\frac{1}{3}$, and $\frac{1}{4}$, written with special signs in hieratic and hieroglyphic notation (see table 2), are also rendered in the usual way by this notation, for example, $\frac{1}{2}$ as $\overline{2}$. This notational system is close to the Egyptian concept and remains the best way of rendering Egyptian fractions.

Following the concept of fractions as inverses of integers, the next step —consequently— would have been to express parts that consist of more than one of these inverses. This was done by (additive) juxtaposition of different inverses; for example, what we write as $\frac{5}{6}$ was written in the Egyptian system as $\overline{2}\,\overline{3}$ ($\frac{1}{2}\,\frac{1}{3}$). The writing of any fractional number thus consisted of one or more different inverses, written according to their size in descending order. Note that this notation enables one to be as accurate as necessary by considering only elements up to a certain size. It also allows for an easy comparison of the size of several fractions; for example, while the immediate comparison of $\frac{5}{8}$ and $\frac{4}{7}$ yields only that they are both a little more than $\frac{1}{2}$, their representation as Egyptian-style fractions, $\overline{2}\,\overline{8}$ and $\overline{2}\,\overline{14}$, allows for immediate comparison.

---

31   See Ritter, "Mathematics in Egypt," p. 631.

8.

Summary

During the Old Kingdom, the social and cultural setting of the scribes was closely connected to the king and his needs. The king was the representative of mankind toward the gods; it was his duty to ensure that life on Earth followed the rules of Maat—the cosmological order of the gods. Only if the king guaranteed the preservation of this order on Earth, which also included carrying out rituals regularly, would the world continue to exist; that is, the sun would continue to rise, harvests would take place, and so on. Experts during the Old Kingdom were to serve their king in this task. Whatever the king ordered was meant to be done to ensure the continuation of Maat on Earth and, therefore, had to be carried out. In return—as is demonstrated for example in the biographical inscription of Weni and Ankh-Meryre-Meryptah—the king would reward his officials for their successes. These rewards were advancements in their careers but also came in the form of various goods, especially expensive objects for their future graves. During the Old Kingdom, the success of experts was judged according to their rank within the royal or temple administration and (linked to this feature) according to their relationship with the king.

The experts had to handle a variety of tasks, among which mathematics must have played an important part. Skillful handling of numbers and numerical data was required to administer the available resources and to carry out the large-scale building projects, expeditions, and other tasks for the king. These activities have left indirect sources of information about mathematics during the Old Kingdom. Various documents attest the development of metrological systems, indispensable prerequisites to control resources. Their handling is also depicted in scenes in tombs of these experts and must have constituted an area of professional pride.

The setting of ancient Egyptian scholarly activities at this time can be traced through the autobiographic inscriptions, also located in the tombs of the officials. Again, various tasks that are listed to represent the work life of an expert suggest that mathematical techniques were part of his training. Explicitly mentioned is the supervision of building projects, which would have required the calculation of building materials but also directing groups of workers in building and executing other tasks, which would have required determining their respective rations and presumably also assessing the amount of work that could be done by a number of people in a specific time.

Apart from this evidence, there are also hints about concepts that are familiar from later sources of Egyptian mathematics. A diagram on a piece of limestone (Cairo JE 50036) found in Saqqara indicates the planning of the construction of a curved structure of some kind (possibly a saddleback construction), and at Meidum a brick wall shows guidelines for the construction of the sloping walls of a mastaba. Both indicate the handling of sloping or curved areas by assessing the difference of vertical height at specific horizontal intervals. The *sqd*, known from mathematical papyri, especially from the pyramid problems of the Rhind mathematical papyrus, was the Egyptian method of describing sloping walls by measuring the vertical difference (in palms) over a horizontal interval of 1 cubit.

MIDDLE KINGDOM

At the end of the Sixth Dynasty, the interplay of several causes (e.g., climatic changes that led to lower levels of inundation followed by famines but also an ever-growing administrative body that swallowed the available resources) caused the breakdown of the Old Kingdom. During the following First Intermediate Period (2160–2055 BCE) the central administration, in which the scribes had developed and executed their mathematics, broke down.[1] Instead of the single authority at the head of the Egyptian culture, two centers of power fought for dominance: The north was ruled by the Heracleopolite Dynasty, based at the entrance to the Fayum. The south was under the government of another dynasty beginning with Antef I, based at Thebes.

On a regional level, local nomarchs, that is, high-level scribes, took over the responsibilities for the care of the population that had formerly been the responsibility of the king. While these nomarchs had used and developed their mathematics to serve the royal administration in providing for and organizing their communities, these abilities were now put to use in their own responsibilities of handling famines and ensuring the execution of basic procedures and securing social order. In their tombs, they proudly recorded their achievements. Thus, Ankhtifi, the nomarch at Hierakonpolis (and a supporter of the Heracleopolites), wrote in the autobiography in his tomb:

> I am the vanguard of men and the rearguard of men. One who finds the solution where it is lacking. A leader of the land through active conduct. Strong in speech, collected in thought, on the day of joining the three nomes. For I am a champion

---

1 A good overview over the First Intermediate Period can be found in Seidlmayer, "First Intermediate Period," 2000.

without peer, who spoke out when the people were silent, on the day of fear when Upper Egypt was silent.[2]

Another official, Merer, from Edfu wrote in his stela:

Never did I hand a person over to a potentate, so that my name might be good with all men. I never lied against any person—an abomination to Anubis. And when fear had arisen in another town, this town was praised. I acquired cattle, acquired people, acquired fields, acquired copper. I nourished my brothers and sisters.

I buried the dead and nourished the living, wherever I went in this drought which had occurred. I closed off all their fields and mounds in town and countryside, not letting their water inundate for someone else, as does a worthy citizen so that his family may swim. When it happened that Upper Egyptian barley was given to the town, I transported it many times. I gave a heap of white Upper Egyptian barley and a heap of *ḥmi*-barley, and measured out for every man according to his wish.[3]

The individual success of a nomarch, as expressed in the autobiographies, was consequently no longer measured through his relation with a superior entity but through his ability to ensure social and economic stability within his own region and through his conduct toward the weak members of its society. In his daily life, then, the same mathematical knowledge that was formerly used to fulfill administrative duties for the king must have played an important role in order to assess, for example, the available grain rations or to organize the work that needed to be done. Through this ability, previously used in the service of the king, the nomarchs were now able to master the new responsibilities that had fallen to them.

However, presumably not all the First Intermediate Period was characterized through these harsh conditions. The height of the Nile inundations seems to have come back to a normal level, and the population must have felt some relief from the lack of monumental royal building projects. An indicator for a certain wealth among the population is the occurrence of a new type of burial goods, made for the Egyptian middle class, which made their first appearance in this period. Before, all but the wealthiest officials were simply buried with their former belongings; now, specifically made grave goods were available, indicating that there seems to have been enough demand for objects of this kind.

---

2   Lichtheim, *Literature I*, p. 86.
3   Ibid., p. 87.

The First Intermediate Period ended with the resolution of the conflict between the two centers of power in form of the reunification of the country by the Eleventh Dynasty Theban king Nebhepetre Mentuhotep II (2055–2004 BCE). The following period of political stability and cultural flowering is known as the Middle Kingdom (Eleventh–Thirteenth Dynasties, ca. 2055–1650 BCE).[4] It lasted until the Eastern Delta emerged during the later Thirteenth Dynasty.[5] Within this period, several phases can be distinguished. During the Eleventh Dynasty, the centralized government was reinstated. Its structure was loosely based on the structure created in the Old Kingdom, where bureaucracy and crown were supported by taxation. The former system of local governors was kept; however, their number seems to have been reduced and they were under closer control of the royal court.[6]

After the Eleventh Dynasty, which had ruled from Thebes, there was a new phase beginning in the Twelfth Dynasty, when the capital was moved to the newly founded town of Itj-tawj ("Seizer of the Two Lands") under its first king Sehetebibra Amenemhat I. Another break can be observed during the reign of King Senusret III through evidence of material culture and textual expression.[7] The kings of the Eleventh Dynasty had sent expeditions to Punt (East Africa) to obtain incense and to Hatnub (alabaster) and Wadi Hammamat (graywacke). Similar expeditions then became more frequent during the Twelfth Dynasty, leading to increasing prosperity for the king as well as middle-class Egyptians. This enabled King Senusret I (1956–1911 BCE) to execute a major construction program, erecting monuments in every main cult site from lower Nubia in the south to Heliopolis and Tanis in the north.[8] The Egyptian borders to Nubia were pushed further to the south and those west of the Walls-of-the-Ruler were consolidated with a series of massive fortifications. Several pharaohs reigning in peace and prosperity followed.

During this time, the Middle Kingdom reached its cultural peak. New types of statuary evolved and the quality of workmanship excelled. Engineers and architects reached new heights of mastery. Literature created during this period reflects the dominance of the royal court.[9] Narratives, prophecies and other literary works depict the king as protector, often against the background of a chaotic kingless period.[10] The literature and the stage of the Egyptian language at that time ("Middle Egyptian") were later considered to be classical.

---

4 Good introductions and further literature can be found in Callender, "Middle Kingdom," and Franke, "Middle Kingdom."
5 Quirke, *Administration*, p. 2.
6 Callender, "Middle Kingdom," p. 152.
7 Quirke, *Administration*, p. 2.
8 Callender, "Middle Kingdom," pp. 161–62.
9 For an excellent analysis of Middle Kingdom literature, see Parkinson, *Poetry and Culture*.
10 See for example the *Tale of Sinuhe*, the *Eloquent Peasant*, and others. Translations of these and more can be found in Lichtheim, *Literature I*; Parkinson, *Sinuhe*; or Simpson, *Literature*.

Sources from the lives of ordinary people are preserved in the settlement of Lahun, the town that provided the support for the funerary complex of King Senusret II. The town accommodated the king's workers and their families. It has been estimated by means of the granaries located throughout the town that up to 5000 people could have lived there. During the excavation of the town by W.M.F. Petrie between 1889 and 1899, several finds of papyri were made. They provide us with sources originating from a nonfunerary context as well as the exceptional case of texts with a known provenance (very often papyri were bought on the antiquities market). It is not surprising, therefore, that they comprise several types of technical texts, including a "calendar" of lucky and unlucky days as well as medical and mathematical handbooks.[11] It is very probable that these mathematical handbooks as well as the other mathematical papyri were used in the training of scribes. We do not have much evidence for the formal organization of education during the Middle Kingdom.[12] No institutionalized schools are known, but temple libraries are attested. Knowledge was presumably passed from father to son or to an individual apprentice. There are no ostraca (the favored material for students' exercises during the New Kingdom), but some writing boards, which may have belonged to student scribes, are extant. They contain model letters, funerary formulas, and name lists.[13]

In addition to the Lahun papyri, a collection of letters dating from the Middle Kingdom deserves to be mentioned. This group of letters allows us insights into the organization of the household of a priest and farmer, a man called Heqanakht. During an absence from his family (due to professional obligations), he wrote to them to make sure his household was properly run. The letters contain information about renting farmland, managing the land, and the distribution of rations within an individual household.[14]

The earliest evidence of mathematical texts originates from the Middle Kingdom (2055–1650 BCE). While it cannot be guaranteed that there were no such texts during earlier periods, it may not be a complete fluke that the extant texts date to this period. From their form and content, it has been assumed that they come from an educational context, and again, while this cannot be guaranteed, the assumption seems reasonable. The mathematical texts teach the mathematics that a scribe would need to use in his daily work, such as calculating volumes of granaries of various shapes, calculating amounts of rations to be distributed, and calculating

11 Parkinson, *Poetry and Culture*, p. 62. A new edition of the Lahun papyri is Collier and Quirke, *UCL Lahun Papyri*.
12 Grajetzki, *Court Officials*, pp. 133 and 146–49. On various aspects of education, see Brunner, *Erziehung*; Brunner, "Schreibunterricht und Schule"; Fischer-Elfert, "Ausbildung"; and Fischer-Elfert, "Schreiber als Lehrer." A possible reference to a school (ˁ.t sbȝ) can be found in a text of a tomb from the First Intermediate Period; see Brunner, *Herakleopolitenzeit*, p. 29.
13 Parkinson, *Poetry and Culture*, p. 236.
14 For an edition of these texts, see Allen, *Heqanakht*.

the amount of produce that has to be delivered by certain professionals. The creation of the genre "mathematical text" thus may have been part of the royal endeavor of the Middle Kingdom to regain and keep control over the country by organizing the mathematical training a scribe would obtain. The extant mathematical texts are a unique source for insights into actual mathematical concepts and techniques.

9.

Mathematical Texts (I): The Mathematical Training of Scribes

The earliest mathematical texts that are extant originate from the Middle Kingdom. While this may well be caused by the vagaries of preservation, it might be that it reflects the actual situation, that is, that mathematical texts of the kind that we have from the Middle Kingdom did not exist in earlier periods. Based on the evidence that we have from the Old Kingdom, this would not mean that mathematical practices did not exist prior to the Middle Kingdom; it would reflect only that they were communicated in a different way. With the reestablishment of central power by the king in the Middle Kingdom also came about a complete new organization of the administrative apparatus that was designed to be much less independent than it had been at the end of the Old Kingdom. And this may well have entailed the organization of teaching mathematics to the future scribes in a centrally organized style, with prescribed problems and their solutions. None of this can be demonstrated by the sources; there are no metatexts that explain the reasoning behind the collections of mathematical problems that are extant. However, the appearance of the genre of mathematical texts fits well within the other information about the new organization of the administrative apparatus.

9.1 EXTANT HIERATIC MATHEMATICAL TEXTS

The term *mathematical text* has taken on a definite meaning among historians of mathematics. As such, mathematical texts are by no means simply all texts that include numbers. Nor are they even all texts that contain mathematical operations. The corpus of mathematical texts has been defined by Eleanor Robson (thus making explicit a previously unspoken tradition since

the beginning of modern research in ancient mathematics) to be restricted to those texts that
have been written about mathematics:

> A distinction must be made between texts which are of mathematical interest (i.e.
> whose contents are of interest to modern historians of mathematics, numeration, or
> metrology . . .) and texts which are primarily about mathematics, i.e., which have
> been written for the purpose of communicating or recording a mathematical tech-
> nique or aiding a mathematical procedure to be carried out.[1]

Only few mathematical texts from ancient Egypt are extant—reflecting once again the
situation of source material being provided mostly by excavations of tombs and temples.
Mathematical texts—having a daily life context and hence being used and discarded in an
urban environment rather than in a cemetery—are thus very rare among the extant material
from ancient Egypt. The choice of writing material used for day-to-day writing such as letters
and accounts—but also for literary, medical, and mathematical texts—was papyrus, which
survives only in absolutely dry conditions. The geography of Egypt provided ideal conditions
for the survival of papyrus in tombs and temples located in the desert; however, of items written and used in the towns along the Nile, little is preserved, mostly by chance, as when these
texts had ended up in the desert.[2]

As for the mathematical texts, so far there are two groups of sources available. The first
group consists of hieratic texts, mostly from the Middle Kingdom. The second group, which
dates from a period more than 1000 years later than the Middle Kingdom, consists of several
papyri and ostraca written in Demotic (a later stage of the Egyptian language).[3] This second
group is described in a later chapter.

The majority of the hieratic mathematical texts come from an educational context. The
largest extant source, the Rhind mathematical papyrus, is likely to have served as the manual
of a teacher. The content of Egyptian mathematical texts can be divided into two categories:
procedure texts (or problem texts) and table texts. Procedure texts present a mathematical
problem, followed by instructions for its solution. Table texts are tabular arrangements of

---

1   Robson, *Mesopotamian Mathematics*, p. 7.
2   A good example of such a chance find are the Heqanakht papyri, a collection of eight letters and accounts
    from a minor official during the Twelfth Dynasty, which were found in the necropolis at Deir el Bahri
    (Allen, *Heqanakht*). For the Middle Kingdom, the excavation of the pyramid town of the pyramid of
    Sesostris II (in the desert) has yielded the largest amount of Middle Kingdom papyri (Collier and Quirke,
    UCL Lahun Papyri).
3   For the development of Egyptian language over time, see Loprieno, *Ancient Egyptian*.

numbers used as aids in calculations. Extant table texts from Egypt include tables for fraction reckoning as well as tables for the conversion of measures. A single source may contain one table or problem only or present a collection of tables and problem texts.

This section gives a brief overview of the extant hieratic mathematical texts dating from the Middle Kingdom and the Second Intermediate Period (the Rhind mathematical papyrus was written during the Second Intermediate Period but states explicitly that it is a copy of an earlier document, which is presumed to originate from the Middle Kingdom). They are mostly written on papyrus. In addition, there is one table written on leather and another table noted on a set of wooden boards. The texts are described here in order of their publication.

9.1.1 Rhind Mathematical Papyrus

The Rhind mathematical papyrus (Rhind papyrus) is the largest extant source text. Today it is kept in the British Museum (London). Being in two pieces, it has the inventory numbers BM 10057 and BM 10058. BM 10057 measures 295.5 cm by 32 cm and BM 10058 is 199.5 cm by 32 cm. The gap between both pieces is assumed to be approximately 18 cm. The diminishment in measurements compared to those given in earlier editions is due to conservation work, which included detaching the papyrus from a permanent backing.[4] The papyrus was bought in Egypt by the British lawyer Alexander Henry Rhind in 1858. After his death, the papyrus was then acquired by the British Museum in 1864.[5] It seems that Alexander Rhind had acquired the mathematical leather roll (see 9.1.5) together with this mathematical papyrus.[6] In 1862, Edwin Smith bought some further fragments of the Rhind papyrus as well as a medical papyrus, today known as the Edwin Smith surgical papyrus. The heirs of Edwin Smith gave the mathematical fragments and the medical papyrus to the New York Historical Society. When the Brooklyn Museum bought the Egyptian collection of the Historical Society in 1949, the Edwin Smith papyrus was handed over to the Academy of Medicine. The fragments of the Rhind papyrus are still in the Brooklyn Museum. Although we cannot be certain of the provenance of any of the three texts, it seems possible that they came from the "library" of one scribe.

The first edition of the Rhind papyrus was published in 1877 by the German Egyptologist August Eisenlohr.[7] Eisenlohr's publication included a hieroglyphic transliteration,

---

4   See the respective note by T.G.H. James in Robins and Shute, *Rhind Mathematical Papyrus*, p. 6.
5   Date cited after Glanville, "Mathematical leather roll," p. 232, while Robins and Shute, *Rhind Mathematical Papyrus*, p. 9, give 1865.
6   Glanville, "Mathematical leather roll," p. 232.
7   Eisenlohr, *Mathematisches Handbuch*.

transcription, (German) translation and a commentary, and a facsimile of the text. Another facsimile was published by the British Museum in 1898. Due to major progress in Egyptology, Eisenlohr's edition was outdated by the end of the nineteenth century, and in 1923, Thomas Eric Peet published a new edition.[8] Peet gave a hieroglyphic transliteration, (English) translation and a commentary. Due to the excellent quality of Peet's translation, this edition has remained the *editio princeps* until today. A third edition was prepared by a group of mathematicians at Brown University. This was published in the years 1927 and 1929.[9] It included, for the first time, photographs (black and white) of the text. The two editions reflect the interests of two academic disciplines in this source, Egyptologists on one hand (who tend to use the excellent edition by Peet) and historians of mathematics on the other hand, who preferred the edition that was made by people they knew. Due to the interest of mathematicians and historians of mathematics in this edition, an abbreviated version of it (without the photographs) was published in 1979.[10] Finally, after the papyrus had been subjected to some cleaning and restoration, a little booklet with color photos of the papyrus was published in 1987.[11]

The original text of the Rhind mathematical papyrus was written during the Second Intermediate Period.[12] However, in its title it is stated that it is a copy of an older text, which is often presumed to be from the Middle Kingdom. The capacity measurements include the quadruple hekat, which is only attested from the end of the Middle Kingdom.[13] Therefore, it is likely that only part of the text was copied from an earlier Twelfth Dynasty manuscript.[14] The definite date *ante quem* is given by the date noted in the text, the year 33 of the Hyksos ruler Apepi (ca. 1550 BCE).

After a broad border at the beginning, which contains the title

Rules for inquiring into nature, and for knowing all that exists, [every] mystery, . . . every secret

---

8   Peet, *Rhind Mathematical Papyrus.*
9   Chace, Bull, Manning and Archibald, *Rhind Mathematical Papyrus.*
10 Chace, *Rhind Mathematical Papyrus.*
11 Robins and Shute, *Rhind Mathematical Papyrus.*
12 For a concise overview of that part of Egyptian history see Booth, *Hyksos Period.*
13 Pommerening, *Hohlmaße*, pp. 120–21: "Gegen Ende des Mittleren Reiches kommt in den Wirtschaftstexten zum einfachen und zweifachen Hekat ein dreifaches und vierfaches hinzu" (however, table 5.2.1.1, which she refers to on p. 128, indicates different writings only for units from the Rhind mathematical papyrus).
14 See Griffith, "Weights and measures," p. 436, note †, Griffith, "Rhind mathematical papyrus," p. 164, Spalinger, "Dates in ancient Egypt," p. 258, and Spalinger, "Rhind mathematical papyrus as a historical document," p. 303. Peet, *Rhind Mathematical Papyrus*, p. 3, maintained that the whole text may be a copy of a Twelfth Dynasty original.

written in red ink in columns, both recto and verso are ruled by seven black horizontal lines that divide the papyrus into six bands.[15] Red ink is used to highlight the beginnings of new sections and as a marker within the texts of problems and tables.

The papyrus contains 64 "problems" as well as several tables. The numbering of the problems today (which includes numbers up to 87) is that introduced by August Eisenlohr in his first edition of the text. However, numbers 7–20 are mere calculations, which are more likely to be associated with the creation of a table, as are the calculations found as number 61. Numbers 80 and 81, likewise, are better referred to as table texts, while 82–84 are (model) accounts. Following these are: in number 85 an unintelligible group of signs,[16] in 86, a fragment of accounts, and in 87, calendrical entries. Two further problems, 59B and 61B, have to be added to the count.

9.1.2 Lahun Mathematical Fragments

The UCL Lahun mathematical fragments are kept in the Petrie Museum (London) with the inventory numbers UC 32107A, UC 32114B, UC 32118B, UC 32134A + B, and UC 32159–32162. They provide the exceptional case of a definite find context. They were found with the major Middle Kingdom papyrus find made by W.M.F. Petrie during his excavation of the town of the pyramid complex of Senusret III. Petrie used the name "Kahun" to denote the town and "Illahun" for the pyramid complex. The papyri found by him have in the past been referred to as "Petrie papyri" and "Kahun papyri." Since there is often no detailed documentation of provenances, the recent publication of the papyri uses the term "Lahun," the simplest form of the local Arabic place name from which Petrie derived his designations.[17]

Most of the fragments were first published by Francis L. Griffith.[18] The publication included black-and-white photographs, hieroglyphic transliteration, transcription, and (English) translation of the more complete texts. Thus UC 32107A, UC 32118B, and UC 32114B were not included in this edition. With the republication of the UCL Lahun papyri in 2002–2006, the mathematical texts were also edited again, including the previously unpublished fragments.[19]

---

15 For the preparation of the text see Spalinger, "Rhind mathematical papyrus as a historical document," pp. 298–302.
16 For a possible interpretation of these signs see Morenz, "Fisch an der Angel."
17 Collier and Quirke, *UCL Lahun Papyri*, vol. 1, p. v.
18 Griffith, *Petrie Papyri*.
19 Imhausen and Ritter, "Mathematical fragments," and Imhausen, "UC 32107A."

All the mathematical fragments date from the Middle Kingdom. Some of them contain only a few signs. Others hold several problems or their workings. One of the fragments (UC 32162) has a vertical title before the text of two problems, which is then followed by an endnote.

9.1.3 Papyrus Berlin 6619

Papyrus Berlin 6619 comprises two bigger fragments measuring 14 cm by 8 cm and 5.5 cm by 7 cm, as well as around fifty small fragments. They are kept at the *Staatliche Museen* in Berlin. Their provenance is unknown. They were bought in Luxor in 1887.[20]

The two largest fragments were published in 1900 and 1902 by Hans Schack-Schackenburg.[21] They contain parts of four problems. However, only three of them are preserved well enough to be restored. Among the extant texts on the biggest fragment are the beginnings of two problems, which are marked with red ink.

9.1.4 Cairo Wooden Boards

The two wooden boards Cairo, CG 25367 and CG 25368, are said to have been found at Akhmim and are today kept in the Egyptian Museum (Cairo). They measure 46.5 cm by 26 cm and 47.5 cm by 25 cm. They are covered on both sides with a layer of polished plaster, on which text has been written in black ink. The two boards were once connected by a string. The first tablet contains the remains of a letter, a list of servants, and mathematical calculations. The second tablet contains another list of servants and further mathematical calculations. The tablets have been dated to the Middle Kingdom by the style of their writing. Photographs and a brief description were included in the Ostraca volume of the *Catalogue général* by Georges Daressy.[22] Their correct interpretation as a table of fractions of the hekat and their corresponding value as a sum of the submultiples of the hekat were given in 1923 by Thomas Eric Peet.[23]

9.1.5 Mathematical Leather Roll

The mathematical leather roll is today kept in the British Museum (London) under the inventory number BM 10250. It measures approximately 40 cm by 24 cm. It is said to have been

---

20 Burkhard and Fischer-Elfert, *Ägyptische Handschriften*, pp. 228–29.
21 Schack-Schackenburg, "Berlin papyrus 6619," and Schack-Schackenburg, "Kleineres fragment." See also Berlev, "Review Simpson."
22 Daressy, *Catalogue général*.
23 Peet, "Arithmetic." For a recent assessment of these boards see Vymazalova, "Wooden tablets."

found and bought by Alexander H. Rhind in 1858, together with the Rhind papyrus. Together, they finally came to the British Museum.

Due to its fragile state, it could be unrolled only in 1926.[24] However, even in its rolled-up state, numbers were visible on the ends of it. Since it had come to the British Museum together with the Rhind papyrus, it was assumed that there was a connection between the two documents. Eisenlohr even had hoped to find in it the original text of which the Rhind papyrus is stated to be a copy. When it finally could be unrolled, it turned out to be no direct relation to the Rhind papyrus. The mathematical leather roll was published in the following year (1927) by Stephen R. K. Glanville.[25]

It contains a table for fraction reckoning, namely, examples of sums of unit fractions that result in a unit fraction (or $\frac{2}{3}$). Those are written in four columns, of which the last two are copies of the first two. Based on the paleography, it is assumed to be contemporary with the Rhind mathematical papyrus.

9.1.6 Moscow Mathematical Papyrus

The Moscow mathematical papyrus (Moscow papyrus) is the second-largest extant source text. While its total length is approximately 5.44 m, its height is only 8 cm. It consists of one big piece (containing thirty-eight columns of text) and nine little fragments. The papyrus was bought by W. Golenischeff between 1892 and 1894. In 1912, it was acquired by the Museum of Fine Arts in Moscow, where it is now held under the inventory number E4676.

The Moscow papyrus was first mentioned in 1894 by Moritz Cantor in the second edition of his *Vorlesungen über die Geschichte der Mathematik I* (Lectures about the history of mathematics, part I). The text was edited in 1930 by Wasili Struve.[26] This edition included black-and-white photos of the text, a hieroglyphic transliteration and a translation and commentary (both in German).

The Moscow papyrus contains twenty-five problems, the first three of which are very badly damaged. It includes the two most intriguing problems of Egyptian mathematics, the calculation of the volume of a truncated pyramid (number 14) and the calculation of a *nb.t* (number 10). This has been interpreted (among other possibilities) as calculation of the surface of a half sphere or the surface of a half cylinder.[27]

---

24 Scott and Hall, "Laboratory notes."
25 Glanville, "Mathematical leather roll."
26 Struve, *Mathematischer Papyrus Moskau.*
27 For a discussion of these options see Hoffmann, "Aufgabe 10." Since then, see also the attempt to identify the object Miatello, "Moscow mathematical papyrus and a tomb model."

9.2 TEACHING MATHEMATICS: MATHEMATICAL PROCEDURE TEXTS

Within the Egyptian mathematical sources we find two types of texts: procedure (or problem) texts and table texts. One source may contain collections of both types of texts; for example; the Rhind papyrus begins after its title with two table texts, followed by several procedure texts.[28]

A procedure text begins by stating a mathematical problem. After the type of problem is announced, some specific data in the form of numerical values are given, thus specifying the problem to one concrete instance or object. This is followed by instructions (the procedure) for its solution. The title, specifications of the problem, and the following instructions are expressed in prose, using no mathematical symbolism. The title (and other parts of the text) may be accentuated by the use of red ink, which is rendered in the translation as SMALL CAPS.

Each instruction usually consists of one arithmetic operation (addition, subtraction, multiplication, division, halving, squaring, extraction of the square root, calculation of the inverse of a number) and the result of it is given. The instructions always use the specific numerical values assigned to the problem. Abstract formulas, or equations with variables, did not exist.

While this general characterization holds for all Egyptian problem texts, each individual source also shows specific (often formal) characteristics. The following examples are taken from the Rhind and Moscow papyri. Problems of the Rhind papyrus often show the way to carry out the instructions given in the procedure. These "written calculations" can be found either directly after the instruction was given, as in the example of papyrus Rhind, number 26, or there may be a separate section for the calculations at the end of the problem.

PAPYRUS RHIND, NUMBER 26

[1] A QUANTITY, ITS $\overline{4}$ TO IT, it results as 15.   *Title and data*

Calculate with 4.   *Procedure (instruction)*

You shall calculate its $\overline{4}$ as 1. Total: 5.

[2] Divide 15 by 5.

---

28   Clagett, *Egyptian Mathematics*, provides an English translation of most of the sources in one handy volume. This book, useful as it is, should be used with care. The Rhind papyrus is given in the facsimile of Chace, Bull, Manning, and Archibald, *Rhind Mathematical Papyrus*, rather than in the form of accurate photographs, which can be found in Robins and Shute, *Rhind Mathematical Papyrus*. The Moscow papyrus is given in photocopies of the plates of Struve, *Mathematischer Papyrus Moskau*, and the transliteration is also that of Struve and contains his mistakes, which have been listed and discussed in several reviews (see Peet, "Review Struve"). The presentation of the mathematical UCL Lahun fragments (listed as Kahun mathematical fragments) is incomplete. For a recent complete publication of those fragments, see Imhausen and Ritter, "Mathematical fragments," and Imhausen, "UC 32107A." For a Czech translation of the hieratic mathematical texts, see Vymazalova, *Hieratické matematické texty*.

[3] \ .	5	*Written calculation*
[4] \ 2	10	
[5] 3 shall result.		*Procedure (intermediate result)*
Multiply 3 by 4.		*Procedure (instruction)*
[6] .	3	*Written calculation*
[7] 2	6	
[8] \4	12	
[9] 12 shall result.		*Procedure (intermediate result)*
[10] .	12	*Verification (numerical)*
[11] $\overline{4}$	3	
Total: 15.		
[12] THE QUANTITY: 12,		*Verification (rhetoric)*
[13] its $\overline{4}$: 3,		
TOTAL: 15.		

Many earlier studies of Egyptian mathematics focused on this type of problem; it is one of the examples where a "modern mathematician" has the impression of knowing at first glance (given that he or she is presented with a translation into a modern language) what is going on "mathematically." This type of problem (other examples can be found in papyrus Rhind, numbers 24, 25, 27, 28, 29, 30, 31, 32, 33 and 34; papyrus Moscow, numbers 19 and 25; papyrus UC 32134; papyrus Berlin 6619, number 1; and a closely related type of problem involving the *ḥqꜣ.t* in the Rhind papyrus, numbers 35–38) was classified as "forerunners" of algebraic equations in one unknown.

This then led to the first major controversy about our understanding of Egyptian mathematics, because some mathematicians insisted the problems were solved using the method of false position, while others took them to be evidence that the Egyptian procedure is "equivalent" to our modern manipulation of algebraic equations.[29] Instead of reading modern rules into this text, let us instead focus on a description of the problem and its formal features. Problem 26 begins by stating the type of problem with its numerical data. The Egyptian technical term for "quantity," *ꜤḥꜤ*, at the beginning of the first line indicates the type of problem as an *ꜤḥꜤ* problem. The announcement of the data is directly followed by the first instruction.

---

29 For a discussion, see Imhausen, "*ꜤḥꜤ*-Aufgaben," and Imhausen, "Algorithmic structure." Even in recent years, this group of problems has not lost the interest of modern researchers; see, for example, Vymazalova, "*ꜤḥꜤ*-problems."

The second instruction, the division of 15 by 5 in the second line, is followed by two lines in which this calculation is performed. This is followed by announcing the result of this step in the fifth line, before the next instruction, the multiplication of 3 by 4. Again, this is carried out in writing in lines 6 to 8, and the result is announced in line 9. This result is also the final result of the problem, which is not indicated explicitly in this example. The solution is followed by a verification in lines 10 to 13, which is carried out twice. The first verification (lines 10 to 11) is performed as a written calculation, that is, purely numerically. The result is subjected to the calculation indicated in the first line. This is followed in lines 12 to 13 by a "rhetoric" verification.

To facilitate a comparison of procedures, Jim Ritter proposed to rewrite them in the form of two kinds of symbolic algorithms, a "lightly symbolized sequence" replacing rhetoric instructions of arithmetical operations with our modern symbols $+$, $-$, $\times$, $\div$, and so on, and a more abstract form in which the given data are represented as D_i and the result of step number (N) by (N).[30] I have used this method for an analysis of all hieratic problem texts, and it has since been used by others as well.[31]

For the problem stated earlier (papyrus Rhind, number 26), this rewriting results in the following symbolic algorithms:

	$\overline{4}$			D_1
	15			D_2
(1)	Inverse of $\overline{4} = 4$		(1)	Inverse of D_1
(2)	$4 \times \overline{4} = 1$		(2)	$(1) \times D_1$
(3)	$1 + 4 = 5$		(3)	$(1) + (2)$
(4)	$15 \div 5 = 3$		(4)	$D_2 \div (3)$
(5)	$3 \times 4 = 12$		(5)	$(4) \times (1)$
(v1)	$4 \times (12) = 3$		(v1)	$D_1 \times (5)$
(v2)	$12 + 3 = 15$		(v2)	$(5) + (v1) = D_2$

---

30 Ritter, "Mathématiques en Égypte et en Mésopotamie" (English translation in Ritter, "Measure for measure"; for the rewriting, see esp. pp. 50–52). The term *algorithm* (derived from the name of the Islamic mathematician Muḥammad ibn-Mūsā al-Khwārizmī (ca. 780–850 CE) who explained in his Kitāb al-jam'wal tafrīq bi ḥisāb al-Hind (*Book on Addition and Subtraction after the Method of the Indians*) how to use a decimal positional system detailing procedures for addition, subtraction, multiplication, division, halving, doubling, and determining square roots (Katz, *History of Mathematics*, p. 225)) has traditionally been associated with a later period of history of mathematics (see Rashed, *Arabic Mathematics*, and Berggren, *Mathematics of Medieval Islam*) and is also strongly associated with modern programming. The Egyptologist Stephan Seidlmayer has pointed out analogies between the style of Egyptian mathematical problems and modern programming languages (Seidlmayer, "Computer im Alten Ägypten"). Note, however, that each of the Egyptian sources consists of more than simply the procedure and that the procedures are sometimes implicitly limited to specific numerical values.

31 See Imhausen, *Algorithmen*, and Melville, "Poles and walls."

This example may serve to sketch some strengths and limitations of this methodology. Its key feature is to preserve the Egyptian style of communicating mathematics as procedures. The rewriting helps to facilitate the comparison of individual procedures and to detect relations between individual problems. My analysis of the whole corpus of ꜥḥꜥ problems revealed that based on their procedures, they can be assigned to several distinct groups.[32] In recent research, Jim Ritter found that the procedures of at least two of these groups can be shown to be variations of a basic procedure.

However, the method is not without limitations, and a modern reader should be aware of them. The rewriting is straightforward as long as a step of the procedure involves a simple arithmetic operation as can be seen from steps (2)–(5). Establishing the first step, however, is not trivial. The instruction found in the problem is read "calculate with 4." The 4 is apparently the inverse of the first datum ($\overline{4}$), but the instruction itself does not mention this relationship. To establish it without doubt in this case, problems 24, 25, and 27 of the Rhind papyrus, in which the same relation can be found between the first datum and a subsequent number, may be used. The instruction does not mention a specific arithmetic operation either. Hence the rewriting of the first step depends on the interpretation of the modern reader. In his rewriting of problem 26 of the Rhind papyrus, Jim Ritter chose a different rewriting for this step.[33] Therefore, the establishing of the symbolic algorithms may not be as straightforward as it seems at first glance; the rewriting includes a certain amount of interpretation, and different readers may arrive at different algorithms (where they "guess" missing instructions).

Certain parts of the problem text itself are not included in the symbolic algorithm, for example, in this problem the written calculations found after the instructions. In some cases (as for example in problems 24, 25, and 27 of the Rhind papyrus) explicit rhetoric instructions may be missing but can be reconstructed based on calculations carried out in writing. If instructions are missing and not indicated by written calculations, the reader can only guess which steps were taken to fill the blank. However, in analyzing the texts, this might be considered a strength of the method, because it alerts the researcher to the fact that certain steps were not recorded. Thus, none of the previously mentioned limitations is a flaw in the methodology itself—but one should be aware of them.

The next example of a problem is taken from the Moscow papyrus, which stretches over three columns (XXVII–XXIX) of the papyrus:

32   See Imhausen, "ꜥḥꜥ-Aufgaben," and Imhausen, "Algorithmic structure."
33   Ritter, "Reading Strasbourg 368," p. 185.

PAPYRUS MOSCOW, NUMBER 14

XXVII

¹ Method of calculating a ▱.	*Title*
² If you are told: a ▱	*Data*
of 6 as (its) height,	
³ to 4 as (its) lower side,	
to 2 as its upper side.	
⁴ You shall square these 4.	*Procedure (instruction)*
16 shall result.	*Procedure (intermediate result)*
⁵ You shall double 4.	*Procedure (instruction)*
8 shall result.	*Procedure (intermediate result)*
⁶ You shall square these 2.	*Procedure (instruction)*
4 shall result.	Procedure (*intermediate result*)

XXVIII

¹ You shall add the 16	*Procedure (instruction)*
² and the 8 and the 4.	
³ 28 shall result.	*Procedure (intermediate result)*
You shall calculate ⁴ 3̄ of 6.	*Procedure (instruction)*
2 shall result.	*Procedure (intermediate result)*
You shall ⁵ calculate 28 times 2.	*Procedure (instruction)*
56 shall result.	*Procedure (intermediate result)*
⁶ Look, 56 is belonging to it.	*Statement of final result*
What has been found by you is correct.	*Statement indicating the end of the problem*

XXIX

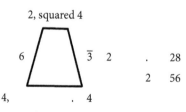

Annotated drawing

2, squared 4

6   3̄ 2   .   28
        2   56

4,   . 4
squared 16   2   8 Total: 28

Number 14 of the Moscow papyrus teaches the calculation of the volume of a truncated pyr-amid. The text begins by stating the problem. The truncated pyramid is not designated by a tech-nical term, but instead it is indicated by a sketch or drawing within the text. This is followed by the numerical data of the object, lower side, upper side, and height. The procedure follows immediately, and, in contrast to number 26 of the Rhind papyrus, it is not interrupted by calculations. One instruction is given, its result announced, and the next instruction follows immediately.[34]

Rewriting this procedure text results in the following symbolic algorithms:

6		D_1	
4		D_2	
2		D_3	
(1)	$4^2 = 16$	(1)	D_2^2
(2)	$4 \times 2 = 8$	(2)	$D_2 \times D_3$
(3)	$2^2 = 4$	(3)	D_3^2
(4)	$16 + 8 + 4 = 28$	(4)	$(1) + (2) + (3)$
(5)	$\overline{3} \times 6 = 2$	(5)	$\overline{3} \times D_1$
(6)	$2 \times 28 = 56$	(6)	$(5) \times (4)$

Note that the rewriting of the text of the procedure in the form of its algorithm obliterates a problem created by a philological nuance of the text. The second step of the procedure is rendered as 4×2, or $D_2 \times D_3$, based on the numerical values in the data and our knowledge of the calculation of the volume of a truncated pyramid. The original instruction reads: "You shall double 4." That is, it is phrased as a specific operation ("doubling"), which does not explicitly make reference to the use of D_3.

The annotated drawing at the end of the text of problem 14 from the Moscow papyrus shows the individual steps of the calculation in its clever annotations. At the top, the length of the upper side is indicated as 2 and the result of squaring it as 4. Likewise, at the bottom of the figure, the length of the base is indicated as 4 and the result of squaring it as 16. On the left side of the figure, the height is indicated as 6; to the right of the figure, a third of the height is noted as 2. The multiplication of the base and upper side of the figure is also noted under the drawing, along with the sum of the squared base and upper side and of the product of base and upper side (total: 28); the final operation, the multiplication of a third of the height and this total is written to the right of the figure.

There have been multiple attempts to explain how the Egyptian formula to calculate the volume of a truncated pyramid could have been achieved; some of these explanations use a clever modification of algebraic formulas—which were not used in Egyptian mathematics.

---

34 Modern researchers have remarked that the procedure used in this problem is very similar to the modern method of calculating the volume of a truncated pyramid using the equation $\frac{1}{3}(a^2 + 2ab + b^2)$.

Other explanations have tried to use practical experiences in form of woodworking. Another possibility would be the use of our knowledge of calculations of volumes of other objects from the hieratic mathematical texts. There are examples of calculations of the volumes of cylinders (e.g., papyrus Rhind, numbers 41 and 42) and a cube (papyrus Rhind, number 44). The underlying strategy for calculating their volumes is a multiplication of base and height. The procedure for calculating the volume of a truncated pyramid may be seen as a variation of that basic concept. Since the shape of the object indicates that a simple multiplication of the base and the height would result in a volume too big (i.e., that of the respective cuboid), a modification is put into place using three different "bases" (the actual base (the square of the lower side), a rectangle made up of lower side and upper side, and the top area (a square of the upper side). To balance the sum of three areas (instead of one), they are then multiplied not by the height but by a third of the height.

9.2.1 The Formal Elements of a Procedure Text

The examples of the Rhind and Moscow papyri just given have shown that a problem text consists of several parts.[35] Necessarily, it should include a title, a presentation of data, and a procedure (consisting of a sequence of instructions) to solve the given problem. Further elements that may (but do not necessarily) constitute part of a procedure text are drawings, calculations performed in writing, and a verification. Due to the fragmentary state of many papyri, sometimes only parts of a problem text are extant, as for example in the papyrus UC 32160.[36] This fragment shows a drawing and three calculations—but no title or procedure. Nevertheless, as in this example, it is often still possible to restore the missing elements.

 Drawings are usually placed at the end of a problem, after the section of instructions. They are not technical drawings in the modern sense, but rather rough sketches illustrating the problem. This distinction is not meant to criticize the ancient scribe for "not having produced proper sketches" but to point to a historiographical problem. Egyptian sketches in general are different from the modern way of making a drawing, and these differences have to be taken into account. Several historians of mathematics have tried to measure the sketches found in Egyptian texts and derive mathematical techniques from them. As Corinna Rossi has pointed out, however, Egyptian drawings are not to scale, and therefore modern attempts to interpret them as such are bound to fail.[37]

---

35 On formal elements of a procedure text cf. Imhausen, "Normative structures."
36 See Imhausen and Ritter, "Mathematical fragments," pp. 84–87.
37 Rossi, *Architecture and Mathematics*, pp. 96–128, esp. pp. 101ff.

Another type of drawing can be found in the title of some of the problems, as, for example, in number 14 of the Moscow papyrus given earlier. The little drawing in the first line of the text represents the object that is the topic of the problem, in this case a truncated pyramid. As Jim Ritter has noticed, this drawing is slightly larger than the hieratic signs of the writing and is, therefore, not likely to be an ideogram.[38] This type of drawing (other examples can be found in numbers 41, 42, and 50 of the Rhind papyrus and number 10 of the Moscow papyrus) is a smaller version of the drawings found at the end of a problem. In some instances (for example in numbers 41 and 50 of the Rhind papyrus), even some given datum is inscribed in it.

The instructions for performing certain arithmetic operations such as multiplying or dividing are sometimes followed by their written execution. A written calculation can be found either directly after the respective instruction (as in number 26 of the Rhind papyrus) or in a separate section at the end of the problem (as in number 14 of the Moscow papyrus). In addition to these fully written calculations, some problems include notices and numbers that can best be described as notes during the (mental?) performance of a calculation. These are mostly found after the instructions. Once the final solution is obtained (by having followed the instructions), a verification of the solution may be carried out. The verification can consist of a set of instructions, written calculations, or both.

The individual sections of procedure text are distinguished by several formal elements. The heading of the problem is usually given as "*tp n jr.t* . . . ": "method of calculating . . . ." Further headings may be used to separate the section of written calculations ("*jr.t mj ḫpr*": "calculating according to the procedure") and the verification ("*tp n sjtj*": "method of verification"). The headings may be written in red ink. The use of these two colors, black and red, in this way within the mathematical documents gives an immediate structure of the text of a problem.

In addition there are further set phrases that serve to structure the text:

- "*gmj.k nfr*": "what has been found by you is correct," which occurs at the end of the instructions in most problems of the Moscow papyrus, and perhaps in one problem each of the papyrus Berlin 6619 and the Lahun papyri.
- "*mj ḏd n.k zš*": "if a scribe tells you," which often introduces the announcement of data.
- "*dj.k rḫ.j*," which introduces the quantity that has to be determined.

---

38   See Imhausen and Ritter, "Mathematical fragments." p. 85, note 7. Jim Ritter has established the distinction between *display drawings* and *in-line drawings*.

Different grammatical forms help further to distinguish the individual parts of a problem text. The headings often use infinitive constructions, while the "instructions" section is dominated by the verb form *sḏm.ḫr.f* and imperatives. The *sḏm.ḫr.f* (the designation is composed of the standard verb in Egyptian paradigms, *sḏm*, "to hear," in the third person singular—expressed by the suffix pronoun .*f*— and the marker of this verb form, *ḫr*) is used to indicate an obligatory consequence from certain conditions.[39] In the mathematical procedure texts, it is used in two functions, namely, to indicate the result of an arithmetical operation and to express an instruction. The first of these uses is obviously destined for this specific verb form. If you are told to add 2 and 3, the result will necessarily be 5. The second use is at first glance less obvious. Why would the *sḏm.ḫr.f* be suitable for the instructions? This can be answered by remembering the type of text we are looking at, a mathematical procedure text, destined to train a scribe to solve mathematical problems. If the given problem is to be solved, the scribe is obliged to follow the procedure laid out in the respective example, and thus the *sḏm.ḫr.f* again is destined to express the specific mood of the text. In English translations, this mood can either be rendered as "shall" or "then + present tense." However, one should stick to one of these options within a single translation. Another verb form used in the instructions is the imperative, again a verb form expressing a strong obligation to the person performing the procedure.

The written calculations are not formulated as prose and can thus easily be distinguished from the rest of the procedure text. They consist of a formally fixed scheme using numbers and checkmarks only, as well as the word *dmd*, "total," to indicate the result.

Generally, the numerical values that appear in the instructions throughout a problem text can be assigned to one of three groups:

- data,
- intermediate results, and
- constants.

Constants are numerical values inherent to a specific type of problem. While the data can vary and thus several individual problems of one type may be obtained, the constants will remain the same. They are characteristic for the respective problem type. For example, even if the values for lower side, upper side, and height in number 14 of the Moscow papyrus were changed, the division of the height by 3 would still remain a step of the procedure. Therefore, 3 is a constant of

---

39 For a detailed discussion of this verb form, see Vernus, *Tense, Mood and Aspect*. For its use in scientific texts see Ritter and Vitrac: "Pensée oriental." For its use in the medical texts, cf. Pommerening, "Lehrtexte," pp. 13–15.

this procedure. Data must be indicated at the beginning of the problem, and intermediate results are values that occur as a result from one step of the procedure. The Egyptian mathematical papyri constitute collections of these procedures. While they all operate on specific numerical values assigned to the individual problems, there is no doubt that the scribe who had learned to solve a problem by following the procedure was able to perform the same procedure using different data.

9.2.2 Collections of Procedure Texts

Within the sources containing a collection of problems (papyrus Rhind, papyrus Moscow, and papyrus UC 32162), the individual problems are separated from one another. The use of red ink for the title and the set phrase *gmj.k nfr*, "what has been found by you is correct" (the latter is used only in the Moscow papyrus), to indicate its end help to mark this separation.

The Rhind papyrus is divided into two sections by two vertical black lines near the beginning of the papyrus. The smaller section at the beginning holds the title of the document, "Method of inquiring into nature, and for knowing all that exists, [every] mystery . . . every secret." The remaining space on the other side of the vertical division is further partitioned by a number of horizontal lines, thus creating several sections. One problem is usually confined within one of these sections, sometimes using several columns. The following problem was then written in the section below, until the bottom of the sheet was reached. The scribe then moved on to the next column. Due to the different lengths of some of the problems, this created several blank areas. In some of these, further mathematical "notes" have been scribbled in. They usually contain an independent mathematical problem. These additional notes do not follow the standard layout of mathematical problems outlined here but are often very much abbreviated. An example of these is number 79 of papyrus Rhind:[40]

PAPYRUS RHIND, NUMBER 79

The contents of a house.	houses: 7
.	2801	cats: 49
2	5602	mice: 343
4	11204	grain: 2301
Total: 19607	kernels: 16807
	Total: 19607

---

40   Another popular problem of this kind is problem 48 of Rhind papyrus, which consists of a drawing of a circle within a square and two calculations without further text. For a recent edition of this problem, see Imhausen, "Egyptian mathematics," pp. 30–31.

This is usually interpreted as the following mathematical problem: There are 7 houses, each house contains 7 cats, each cat has eaten 7 mice, each mouse has eaten 7 halms of grain, each halm of grain contained 7 kernels. What is the sum of all of these? The complete text of the source consists of the title followed by a calculation (left column of the preceding translation) and a list of the individual items and their numbers (right column of the preceding translation). The calculation is the multiplication of 7 times 2801, which can be explained as an alternative method of determining the requested total.[41]

The text of the Moscow papyrus is also written in columns, each column having a width of approximately 10 cm. Because the height of the Moscow papyrus is much smaller than that of the Rhind papyrus, each column contains only up to eight lines of text. One problem comprises between one and six of these columns. To mark the beginning of a new problem, the scribe usually began with a new column (the only exception of this is number 13, which begins in the fourth line of the column in which number 12 ends). Red ink is used for the headings up to number 16, and 15 out of the 22 problems conclude with *gmj.k nfr* (the first two problems are too damaged and not included in this count, and number 4 is destroyed in the bottom part of its instructions and therefore is not counted here either).

Papyrus UC 32162—although containing only two problems—provides us with the only example of a mathematical handbook that has a title and an explicit end notice. The title is much shorter than that of the Rhind papyrus and simply reads "Method of calculating matters of accounts." It is written vertically in red ink. This is followed by two problems written in three columns. Each of the problems starts in a new column with a heading written in red ink. At the end of the last column we find the notice *jw.f pw* ("it is finished")[42] to indicate the end of the composition.

---

41 This problem is another example where modern mathematics has obscured the historical judgment of previous scholars. Today, we would interpret this problem as an example for a geometric series, which we calculate using the formula

$$S = a\frac{r^n - 1}{r - 1},$$

which for the numerical values of this problem would be

$$S = 7\frac{7^5 - 1}{7 - 1} = 7\frac{16806}{6} = 7 \times 2801.$$

Based on the possibility that the formula could be calculated in a way that results in the multiplication indicated in the text of Rhind papyrus, no. 79, it has been claimed that the Egyptian scribe knew of the geometric series and this formula to calculate it. Using two of the criteria that were set up by Eleanor Robson (Robson, "Plimpton 322," p. 176) to evaluate an ancient mathematical text, that is, *historical sensitivity* and *cultural consistency*, speculations of this kind can for now be excluded. There is no evidence whatsoever within the entire hieratic corpus of the use of a procedure "equivalent" to this formula.

42 For further attestations of this phrase or its fuller version, "it is finished from beginning to end as found in writing," see Quirke, "Archive," p. 380.

9.3 TYPES OF MATHEMATICAL PROBLEMS

Although the titles of the individual problems indicate that specific problems were thought to belong to the same class, no general Egyptian classification of mathematical problems is extant (an onomasticon with a list of problem names would have been very informative for Egyptian categories of mathematics, but none has been found, and it is doubtful if one existed). Due to the fragmentary character of many sources, only the Rhind and Moscow papyri can at least theoretically be used in order to try to establish Egyptian classifications. The problems found within the mathematical papyri can be assigned to several groups according to their content. A basic distinction (from a modern point of view) can be made between problems with and without a practical background (or applied mathematical problems and pure mathematical problems).[43] The problems without a practical background teach basic mathematical techniques, such as procedures to determine an unknown quantity from a given relation (ꜥḥꜥ problems) or the calculation of areas. The problems with a practical background are formulated as a daily life situation in which mathematical knowledge is required, for example, the determination of the content of a granary or the distribution of rations. Most of these practical problems can be interpreted as "real" practical problems, that is, those that are likely to have occurred in the work life of a scribe. However, some clearly do not fall into this category, as, for example problem 79 from the Rhind papyrus discussed earlier. In this problem a pseudo-practical background is used to phrase a mathematical problem. Another example of a problem that does not appear to be a straightforward practical instruction for dealing with a daily life situation is number 67 of the Rhind papyrus:

PAPYRUS RHIND, NUMBER 67

METHOD OF CALCULATING THE PRODUCE (*bꜣkw*) OF A HERDSMAN. This herdsman has come to the counting with 70 oxen. This accountant said about the cattle of this herdsman: "How small is this number of cattle that you have brought. Where is the large number of your cattle?" This herdsman said to him: "I have brought to you $\overline{\overline{3}}$ of $\overline{3}$ of the cattle that you had entrusted to me. Calculate for me! You will find me being complete. . . . "

The situation itself, the counting of the produce of cattle, is something a scribe could have faced in his daily work. This is not only depicted in several tombs and models, but it is also

---

43   This distinction cannot be seen at first glance from the Rhind papyrus, however, where we find distributions of a number of loaves among the first problems, followed by a series of purely arithmetical problems (including the ꜥḥꜥ-problems), then again by further problems with practical background.

documented in a papyrus among the fragments from Lahun (papyrus UC 32168).[44] Neverthe-less, the phrasing of this problem differs from the usual practical problems. Instead of a prob-lem and the procedure used to solve it, a little episode is given, in which a herdsman comes to the accounting of the cattle and the scribe questions the amount of cattle that he brought with him. In return, the herdsman indicates the number he has brought in form of a quota ($\overline{\overline{3}}$ of $\overline{3}$), encourages the scribe to calculate it "for him," and predicts that the scribe will find that he has brought the required number. The text of the problem does not indicate the original number of cattle that had been entrusted to the scribe. It is followed by several calculations. First $\overline{\overline{3}}$ of $\overline{3}$ is determined as an Egyptian fraction ($\overline{6}\ \overline{18}$). Then the inverse of this fraction is determined ($4\overline{2}$). This is followed by the instruction to multiply 70 (i.e., the number of cattle the herdsman brought to the accounting) by $4\overline{2}$ (315). This result is described as "these are what had been entrusted to him." Finally, the indicated quota ($\overline{\overline{3}}$ of $\overline{3}$) of this is calculated, and the result is denoted as "these are what were brought to him."

The practical problems can be further divided into two groups according to the area of their practical background. There are problems with an administrative background (e.g., the distribution of rations, the volume of granaries, problems concerning amounts of flour and bread and beer), which constitute the majority of the extant problem texts. But there is also a small group of problems relating to architecture (e.g., the volume of a truncated pyramid and problems about base, height, and inclination of pyramids). The distribution of these problems within the individual sources should help to learn about an Egyptian classification system for mathematical problems. However, the twenty-five problems of the Moscow papyrus seem to have been written down in no specific order. This can best be seen from the distribution of the bread and beer problems that comprise the largest single problem type in this papyrus. The eleven problems can be found as numbers 5, 8, 9, 12, 13, 15, 16, 20, 21, 22, and 24—well spread out over the whole text. The easiest of them is number 15, while numbers 5, 8, 9, 13, and 22 constitute the more advanced representatives. Two of these were written down twice: number 5, which we find again as number 8, and number 9, which has a copy in number 13. Note that these copies are not exact duplicates but rather a second attempt at the same prob-lem. They differ in the detail of the notation of the procedure. Number 8 appears less abbrevi-ated than number 5, and number 9 is more detailed than number 13. Problems involving the

---

44 Depictions in Middle Kingdom tombs can be found in Newberry, *Beni Hassan I*, pl. XIII (tomb 2), Newberry, *Beni Hassan II*, pl. XVII (tomb 17), Davies, *Deir el Gebrâwi*, pl. VII, Blackman, *Meir IV*, pl. XVI, and Davies, *Antefoker*, pl. XIII; the most detailed and best preserved model is depicted in Winlock, *Models of Daily Life*, pl. 13. For the Lahun fragment UC 32168 papyrus, see Collier and Quirke, *UCL Lahun Papyri*, vol. 3, pp. 56–61.

manipulation of an unknown quantity, the so-called ꜥḥꜥ problems, are found as numbers 19 and 25, and the three problems involving geometrical shapes with a right angle are numbers 6, 7, and 17. The problems of the Rhind papyrus, on the other hand, appear to have been arranged in a carefully thought-out order. Groups of problems include ꜥḥꜥ problems (numbers 24–34), directly followed by a similar group of problems involving the ḥqꜣ.t measure (numbers 35–38). Further groups are problems calculating the volume of a granary (numbers 41–46), the calculation of areas of different shapes (numbers 48–55), problems concerning pyramids (numbers 56–60), and bread and beer problems (numbers 69–78). The only problems that seem to be spread out are problems concerned with the determination of rations (found as numbers 1–6, 39–40, and 63–66). However, this may be explained by the differences of the individual subgroups.[45]

This organization of collections of information according to subject is also used in other types of Egyptian texts, for example in onomastica or medical handbooks. Onomastica are catalogues in which groups of things are listed according to their kind. The Middle Kingdom Ramesseum Onomasticon contains plant names, liquids, birds, fishes, quadrupeds, and so on, sometimes preceded by a classificatory heading.[46] The medical handbook found as the Edwin Smith surgical papyrus contains instructions for dealing with wounds of the head (cases 1–7), wounds of the nose (cases 11–14), wounds in the maxillary region (cases 15–17), injuries of the temple (cases 18–22), wounds of the cervical vertebra (cases 29–33), and others arranged in groups (as can be seen from the numbers of cases) according to body parts.[47]

---

45 See my discussion of these problems in Imhausen, "Calculating the daily bread."
46 Gardiner, *Onomastica*.
47 See Breasted, *Edwin Smith Surgical Papyrus*, or the recent translation in Allen, *Medicine*, pp.70–115.

10.

Foundation of Mathematics

At the very heart of a mathematical culture that develops mathematical techniques (e.g., to determine an unknown number from a given relation, the size of a geometric figure of given dimensions, the volume of a granary, or the amount of produce to be delivered by a worker) are basic arithmetical operations that allow the numerate scribe to skillfully manipulate given numbers to determine, for example, their product or their sum. The development of arithmetical techniques necessarily comes after the invention of a number system, will be influenced by it, and in turn will deeply influence any further mathematical techniques that are built on their foundation.[1] The number system and number notation will influence which types of calculations will be easy and which will be difficult to be carried out. The nonpositional decimal system of Egypt renders multiplications by 10 easy, because the scribe needs only to exchange the type of sign for that of the next power. On the other hand, calculations that involve fractions are likely to be technically demanding in a system that relies on the use of series of inverses.

10.1 ARITHMETIC TECHNIQUES

The mathematical texts contain specific expressions for a variety of arithmetic operations. These *termini technici* were often derived from daily life expressions picturing the mathematical operation. Thus, the term for addition is *wȝḥ … ḥr …* "to lay … on …." For examples of

---

1 This has been demonstrated impressively in Ritter, "Mathématiques en Égypte et en Mésopotamie," English translation in Ritter, "Measure for measure," where the calculations of the volume of a granary in Mesopotamia and Egypt are compared.

TABLE 4: Egyptian expressions (examples) for addition, subtraction, multiplication, and division

Egyptian expression	Literal translation	Operation	Attestations within the individual mathematical papyri
wꜣḥ … ḥr …	to lay on	to add	R(6), M(1)
rdj ḥr	to give upon		R(2)
ḫbj	to reduce	to subtract	R(8), L(1)
jrj … r zp …	to make … times …	to multiply	R(17), M(16), L(6), B(1)
wꜣḥ-tp m … r zp …	to bow the head with … times …		R(20)
jrj … r gm.t …	to make … to find …	to divide	R(5), M(17), L(2), B(1)
wꜣḥ-tp m … r gm.t …	to bow the head with … to find …		R(15), L(1)
njs … ḫnt/ḫft …	to call … before …		R(7), M(3)

Note: Abbreviations are R = Rhind Papyrus, M = Moscow Papyrus, L = Lahun Mathematical Fragments and B = Papyrus Berlin 6619. Numbers in parentheses indicate the number of attestations within the respective source.

expressions for basic arithmetic operations, see table 4. The terms listed in this table are those most frequently used within the hieratic mathematical texts. The terms listed in the table serve to indicate parallels and variations among the various sources.

A first observation can immediately be made, namely, that there are specific words designating the major mathematical operations, which, at first glance, occur throughout the corpus of mathematical texts. Absences within individual papyri may be as likely to originate from the vagaries of preservation as from the specific use of individual terms within one source.

A special role within the mathematical texts is held by the verb *jrj* (basic meaning in Egyptian is "to make, to do"). It is found either absolutely in the meaning of "to calculate" (i.e., the basic action of what one does within the mathematical texts) or in combination with other words to express specific mathematical operations (for examples from the Moscow mathematical papyrus, where this usage of *jrj* is quite frequent, see table 5).

In addition to the instructions to perform these operations, the procedure texts sometimes also contain the actual performance, the carrying out of an arithmetic operation. These are mostly multiplications and divisions, as well as few examples of additions of fractions. Multiplications and divisions were carried out in a similar way, using one of several possible techniques, most commonly those of doubling and decupling. The choice of the actual technique depended on the numerical values involved.

TABLE 5: Expressions with *jrj* in the Moscow mathematical papyrus

expression	translation	attestations
jrj	calculate	M-05-04, M-05-08, M-07-05, M-08-05, M-09-07, M-09-16, M-09-17, M-09-18, M-10-05, M-10-08, M-11-05, M-11-06, M-13-07, M-14-09, M-16-05, M-16-08, M-17-07, M-18-03(?), M-19-05, M-20-04, M-20-05, M-21-03, M-21-04, M-22-06, M-22-09, M-23-05, M-24-09
jrj zp	multiply	M-05-11, M-06-04, M-07-03, M-08-09, M-09-20, M-10-12, M-11-08, M-13-11, M-14-11, M-15-05, M-16-06, M-17-06, M-18-05, M-19-01, M-24-05, M-25-01
jrj qnb.t	calculate the root	M-06-05, M-17-06
jrj m znn	square	M-14-04, M-14-06
jrj gs	halve	M-08-08
jrj ḏ3.t	calculate the remainder	M-09-08, M-10-07, M-10-10, M-13-08, M-22-07
jrj ꜥ3	calculate the difference	M-19-03
jrj dmḏ	calculate the sum	M-21-05, M-21-06, M-25-02
jrj r gm.t	divide	(M-05-06), M-06-03, M-08-06, M-09-24, M-11-07, M-12-06, M-12-09, M-13-13, M-16-07, M-17-05, M-19-04, M-20-03, M-22-08, M-23-06, M-24-04, M-24-07, M-25-03

Note: Occurrences are cited by number of the problem in the Moscow papyrus and line, e.g., M-06-05 indicates problem 6, line 5 of the Moscow mathematical papyrus.

10.1.1 Multiplication of Integers

The multiplication of two integers is designated in Egyptian as *wꜣḥ-tp m . . . r zp . . .* or *jrj m . . . r zp . . .* , which means literally "take up . . . times . . . ."[2] This reflects the formal scheme in which the multiplication is carried out. The first example, the multiplication of 3 and 4, is taken from number 26 of the Rhind papyrus. The instruction given in the procedure is *wꜣḥ-tp m 3 zp 4* (calculate with 3 four times). This is carried out in the following way:

.	3	*Initialization*
2	6	*Doubling*
\4	12	*Doubling*

---

2   On the usage of *zp* in expressions for multiplication, see Neugebauer, "Konstruktion von *sp*."

The layout of the performance of this operation comprises two columns. The first line, the initialization of the process, shows a dot in the first column, and the number that is affected by the calculation (in this example, 3) in the second column. The technique chosen here consists of repeated doubling. The dot in the first column counts as 1. Thus, the second line has 2 in the first column and 6 in the second column. This is continued. In order to obtain the result of the required multiplication, there is a permanent watch over the first column while the doubling progresses. The scribe executing the multiplication tries to find entries in this first column that add up to the multiplier, in our case 4. These lines are then marked with a dash or checkmark (\). In our example, it is only the third line that is marked. The last step of the multiplication consists of adding the entries in the second column of these marked lines. The addition is noted below the scheme as *dmd* (total), and this total is the result of the multiplication. In this example, it is only the third line that needs to be considered, and it immediately gives the result (12). Therefore, the scribe did not bother to write out the total separately.

Another example of the same technique can be found in number 52 of the Rhind papyrus. The performance of the multiplication is as follows:

\.	2000	*Initialization*
2	4000	*Doubling*
\ 4	8000	*Doubling*

Total: 10,000

Here, 2000 shall be multiplied 5 times. The technique used is the same as in the previous example. After the initialization, the carrying out continues by repeated doubling. As 2000 is to be multiplied 5 times, the first and third lines are marked (\) and then added to obtain the result of the multiplication.

Another technique consists of decupling, as can be seen from the multiplication 75 × 20 of problem 44 of the Rhind papyrus:

.	75	*Initialization*
10	750	*Decupling*
\ 20	1500	*Doubling*

Again, the multiplication begins with an initialization. Instead of doubling this first line, the scribe directly moves on to 10 in the second line (decupling). This is followed by doubling in the third line. As in our first example, the multiplier (20) can directly be found in one line, and thus no further addition is needed.

The last example is taken from the Lahun fragment UC 32160. The multiplication carried out looks as follows:

\\.	16	*Initialization*
\\ 10	160	*Decupling*
\\ 5	80	*Halving*
Total: 256		

Thus we find here the multiplication of 16 times 16. After the initialization and the decupling, the third step consists of halving the previous line. The entries in the first column (. = 1, 10, and 5) now add up to the multiplicative factor 16. All three lines get marked (\\), and the entries in the second column are added to obtain the result of the multiplication (256).

10.1.2 Division of Integers

Egyptian expressions to execute a division include *njs … ḥnt …* , literally "call … before … ," and *jrj/wȝḥ-tp m … r gm.t …* , or "calculate with … to find …." The similarity to multiplications (*jrj/wȝḥ-tp m … (r) zp …*) of this last *terminus technicus* reflects the similarities in carrying out these operations. The formal scheme again consists of two columns, and we find the same techniques as in multiplication employed within this scheme. The following example, found in number 26 of the Rhind papyrus, is noted directly after its respective instruction *wȝḥ-tp m 5 r gm.t* 15, "divide 15 by 5."

\\.	5	*Initialization*
\\ 2	10	*Doubling*

Again, the carrying out begins with an initialization by putting a dot in the first column and the number that is operated upon (the divisor) in the second column. This is doubled in the second line (as in multiplications, the dot counts as 1). While doubling, the second column is closely watched for elements that add up to the dividend. In our example, the first two lines happen to add up to the dividend (15). Both get checkmarks. In order to obtain the result of the division, the marked lines of the first column have to be added.

The "total," which is sometimes given at the end of the written performance of a division, can either be the sum of the relevant entries of the second column (the dividend)[3] or the sum of the relevant entries of the first column (the result of the division).[4] In this example, the carrying out of the division has been placed directly after its respective instruction. It is followed by indicating its result in the usual style found in procedures: *ḫpr.ḫr* 3, or "3 shall result."

The technique of decupling combined with doubling can be seen in number 69 of the Rhind papyrus. The instruction in the procedure is *jrj.ḫr=k wȝḥ-tp m 80 r gm.t* 1120, "you shall divide 1120 by 80." The division is carried out as follows:

.	80	*Initialization*
\ 10	800	*Decupling*
2	160	*Doubling*
\ 4	320	*Doubling*
Total: 1120		

After the initialization, the next step consists of decupling. This is followed by two steps of doubling, the first of which refers back to the first line as reference. The second and fourth lines are marked since they add up to 1120 (the dividend), which is then indicated as the total. The result of the division (obtained by adding the checkmarked lines of the first column) is not explicitly noted.

10.2 FRACTION RECKONING

The Egyptian notation of fractions as unit fractions (or, better, fractions "without" numerator) resulted in the manipulation of a series of such fractions. Thus, the domain of fraction reckoning appears to be rather laborious and sometimes cumbersome. In order to enable a smooth handling of fractions, tables that listed transformations needed repeatedly (such as the doubling of fractions or the addition of several fractions) had to be created. These tables and their use are discussed in detail in a following section. The next section introduces techniques used in arithmetic operations that involved fractions.

---

3   As, for example, in papyrus Rhind, numbers 21, 22, 27, 31, 35, 67, and 69.
4   Ibid., numbers 30, 32, 34 and 76.

10.2.1 Multiplications and Divisions Involving Fractions

Fractions may appear in both columns of the written multiplication. If the number that shall be operated upon (which is then found in the second column of the first line (initialization) is to be multiplied by a number involving fractions, the integer part of the multiplication is carried out first in the usual way. This is followed by the fractional parts. Fractional parts usually refer back to the initialization and not the line immediately preceding them. The following example is taken from number 69 of the Rhind papyrus. The multiplication carried out is $3\,\overline{2} \times 320$.

\.	320	*Initialization*
\2	640	*Doubling*
\$\overline{2}$	160	*Fractional part (halving)*
Total: 1120		

Here again, several techniques can be used. The fractional parts often progress along a series of halving, starting with a suitable fraction—for example, $\overline{3}$, as in numbers 29 and 43 of the Rhind papyrus, $\overline{2}$, as in numbers 53, 58, and 59 of the Rhind papyrus, or $\overline{10}$, as in numbers 41, 42, and 44 of the Rhind papyrus. The respective fraction is either reached during this series of halving or can be built up from several fractions.

If the fractions involved are rather simple, like $\overline{3}$, the fractional part may also be noted directly, as in numbers 29 ($\overline{4}, \overline{10}$) and 35 ($\overline{3}$) of the Rhind papyrus. In papyrus UC 32160 we find the following calculation:

\.	256	
2	512	
\4	1024	
\$\overline{3}$	85 $\overline{3}$	
Total: 1365 $\overline{3}$		

The multiplication to be carried out is $5\,\overline{3} \times 256$. As usual, the procedure starts with the initialization, followed by the integer part. This time, the technique of successive doubling is used. When this is completed, the fractional part might have begun (following the schemes found elsewhere) with $\overline{\overline{3}}$ (and the intention to continue to $\overline{3}$ from there). Instead, the value of $\overline{3}$ (85 $\overline{3}$) is directly written in the second column, and the total is calculated. The detailed notation would have been as follows:

\.	256
2	512
\4	1024
$\overline{\overline{3}}$	170 $\overline{\overline{3}}$
\3	85 $\overline{3}$

Total: 1365 $\overline{3}$

10.2.2 Using Auxiliary Numbers

Divisions involving fractions often include quickly very small fractions (i.e., fractions with a large denominator). In order to carry out this kind of division more easily, at first only fractional values up to a certain size are taken into account. The remaining fractions with larger denominators are then handled separately, often using auxiliary numbers.[5] An example of this strategy can be seen in number 31 of the Rhind papyrus. The division 33 ÷ 1 $\overline{3}$ $\overline{2}$ $\overline{7}$ is carried out as follows:

1	1 $\overline{3}$ $\overline{2}$	$\overline{7}$
\2	4 $\overline{3}$ $\overline{4}$	$\overline{28}$
\4	9 $\overline{6}$	$\overline{14}$
\8	18 $\overline{3}$	$\overline{7}$
$\overline{2}$	$\overline{2}$ $\overline{3}$ $\overline{4}$	$\overline{14}$
\$\overline{4}$	$\overline{4}$ $\overline{6}$	$\overline{8}$ $\overline{28}$

Total: 32 $\overline{2}$

Remainder: $\overline{2}$

The division is carried out in the usual way, using the technique of successive doubling. Once the doubling has reached 8 in the first column, the value in the second column is 18 $\overline{3}$ $\overline{7}$, and thus the end of the doubling is reached (because the next step would result in a number larger than the dividend). A second series of consecutive halving is then started. As usual, the scribe tries to add suitable lines of the second column to reach the dividend. This is rather complicated, due to the presence of fractions. In order to make it easier, only fractions up to $\overline{6}$ are considered at first. Thus, the total of the marked lines is given as 32 $\overline{2}$. The fractions $\overline{7}$, $\overline{8}$,

---

5   Further examples of the use of auxiliary numbers are discussed in Neugebauer, *Vorlesungen*, pp. 137–47.

$\overline{14}$, and $\overline{28}$ are not included in this addition. These are considered separately, in order to find out how much of the remaining $\overline{2}$ they add up to. (Note that this is a typical *skm* problem.) For this, the auxiliary number $\overline{42}$ is used.

$\overline{7}$	$\overline{8}$	$\overline{14}$	$\overline{28}$	$\overline{28}$	
6	$5\,\overline{4}$	3	$1\,\overline{2}$	$1\,\overline{2}$	$17\,\overline{4}$

(Remainder) $3\,\overline{2}\,\overline{4}$ (since) $\overline{2}$ (is) 21.

The fractions, transformed with the auxiliary number $\overline{42}$, add up to $17\,\overline{4}$. The fraction $\overline{2}$ transformed with the same auxiliary number is 21, and thus the remainder (with regard to the same auxiliary number) is $21 - 17\,\overline{4}$, or $3\,\overline{2}\,\overline{4}$.

The divisor is now transformed with the inverse of the auxiliary number:

1	42
$\overline{3}$	28
$\overline{2}$	21
$\overline{7}$	6

Total: 97 (*sic*! text 99)

To determine the remainder, the division $3\,\overline{2}\,\overline{4} \div 97$ has to be carried out. This is done as a multiplication of the inverse $\overline{97} \times 3\,\overline{2}\,\overline{4}$:

$\backslash\,\overline{97}$	1
$\backslash\,\overline{56}\,\overline{679}\,\overline{776}$	2
$\backslash\,\overline{194}$	$\overline{2}$
$\backslash\,\overline{388}$	$\overline{4}$

The final result of the division is not written out. Instead the total, 33, is noted.

10.3 TABLES FOR FRACTION RECKONING

In order to simplify operations that were not easily performed mentally, tables were created in which the result of the operations could be looked up. Tables found in the hieratic mathematical texts include tables for fraction reckoning and metrological conversions.[6]

---

6   For a list of extant division tables, see Fowler, *Mathematics of Plato's Academy*, p. 269.

10.3.1 2 ÷ n *Table*

As the earlier examples of multiplications and divisions document, doubling a number was a frequently recurring task to be carried out while performing these operations. This is mostly straightforward and easy if only integers are involved (which may have contributed to the development of the Egyptian technique of multiplying and dividing). But it is at first glance rather complicated if fractions do occur. However, fractions with an even denominator are easily doubled by halving the denominator, which results naturally in another Egyptian fraction. Fractions with an odd denominator, on the other hand, are not straightforwardly doubled—remember that the result must be an Egyptian fraction, that is, a unit fraction (or $\overline{\overline{3}}$) or a sum of different unit fractions. If $\overline{5}$ is to be doubled, some thinking may lead to the result that $2 \div 5 = \overline{3}\,\overline{15}$.

However, for an operation to be performed as often as doubling, it is too cumbersome to find a solution again each time the problem is encountered. Thus a table is needed from which the result can readily be taken and used. The $2 \div n$ table is probably the most important table in Egyptian mathematics. Two copies of this table are extant. One of them can be found at the beginning of the Rhind papyrus, and the other is found on one of the Lahun fragments (UC 32159). The table of the Rhind papyrus includes representations of the divisions $2 \div n$ for odd n from 3 to 101. The Lahun fragment holds the same for odd n from 3 to 21. While the Lahun table contains only numerical symbols, the table found in the Rhind papyrus also includes the instruction for the operation to be tabled out at the top of each column. Note that it is expressed as a division, *njs 2 ḥnt* . . . , "divide 2 by . . . ," rather than as the doubling of a unit fraction.

From a modern mathematical point of view, there are several possible representations of $2 \div n$ that would qualify as Egyptian fractions. Nevertheless, it seems that Egyptian mathematics used only one of them. The two extant tables of the Rhind papyrus and the Lahun fragment both show the very same representations, and whenever we find a doubling of an odd unit fraction, it is this representation from the $2 \div n$ table that is used.

Figure 15 shows the $2 \div n$ table found in the Lahun fragment UC 32159 and its transliteration. (Remember that the hieratic text is to be read from right to left, while the transliteration is to be read from left to right). The table is organized as follows: The numbers are arranged in two columns. The first column holds the divisor n (only the first line contains both dividend and divisor). In the second column, a fraction of the divisor and its value (as Egyptian fraction) alternate. Thus one reads, in the second line, the number 5 in the first column—it is $2 \div 5$ that will be expressed as an Egyptian fraction. In the second column of that line, we find $\overline{3}$, 1 $\overline{3}$, $\overline{15}$, and $\overline{3}$. This has to be read as $\overline{3}$ of 5 equals 1 $\overline{3}$, and $\overline{15}$ of 5 equals $\overline{3}$. Since 1 $\overline{3}$ + $\overline{3}$ = 2, the Egyptian fraction to represent $2 \div 5$ is $\overline{3}\,\overline{15}$.

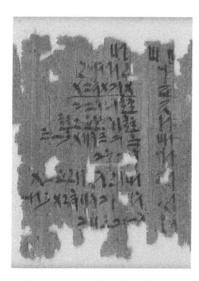

2 3	$\overline{\overline{3}}$ 2
5	$\overline{3}$ 1 $\overline{3}$ $\overline{15}$ 3
7	$\overline{4}$ 1 $\overline{2}$ $\overline{4}$ $\overline{28}$ 4
9	$\overline{6}$ 1 $\overline{2}$ $\overline{18}$ 2
11	$\overline{6}$ 1 $\overline{3}$ $\overline{6}$ $\overline{66}$ 6
13	$\overline{8}$ 1 $\overline{2}$ $\overline{8}$ $\overline{52}$ 4 $\overline{104}$ 8
15	$\overline{10}$ 1 $\overline{2}$ $\overline{30}$ 2
17	$\overline{12}$ 1 $\overline{3}$ $\overline{12}$ $\overline{51}$ 3 $\overline{69}$ 4
19	$\overline{12}$ 1 $\overline{2}$ $\overline{12}$ $\overline{75}$ 4 $\overline{114}$ [6]
21	$\overline{14}$ 1 $\overline{2}$ $\overline{42}$ 2

FIGURE 15: 2 ÷ *n* table of the Lahun fragments (Courtesy of the Petrie Museum of Egyptian Archaeology, UCL)

The 2 ÷ *n* table of the Rhind papyrus differs slightly in several formal aspects. The two types of numbers of the second column, the fractions of the divisor and their values, are distinguished by the use of red and black ink (the fractions of the divisor are written in red, their values in black). Thus it is easier to differentiate between the two types of numbers. In addition, after the listing of each representation of 2 ÷ *n*, there is a verification of the given result. This verification is executed by a multiplication of the result by the respective value *n*. Figure 16 shows the representations of the 2 ÷ *n* table (without verification) as found in the Rhind papyrus. Red ink is rendered as bold print.

The extensive list of values for 2 ÷ *n* makes the modern reader wonder how this specific representation was chosen (among all possible ones)—or, more accurately, why do we have exactly those representations? During the first half of the twentieth century, two historians of mathematics, Otto Neugebauer and Kurt Vogel, tried to answer this question. The dissertation of Otto Neugebauer about Egyptian fraction reckoning was published in 1926.[7] Kurt Vogel finished his thesis about the 2 ÷ *n* table of the Rhind papyrus in 1929.[8] Both of them analyze the representations we find in the 2 ÷ *n* table through modern mathematical formulas.

In the 1970s, Richard Gillings attempted to explain the "rules" behind the selection of these specific representations found in the 2 ÷ *n* table.[9] However, his set of rules are, instead,

7   Neugebauer, *Bruchrechnung*.
8   Vogel, *Grundlagen der ägyptischen Arithmetik*.
9   Gillings, *Mathematics in the Time of the Pharaohs*, pp. 45–70.

2 ÷ 3	$\overline{\overline{3}}$ 2	53	$\overline{30}$ 1 $\overline{3}$ $\overline{10}$ $\overline{318}$ 6 $\overline{795}$ $\overline{15}$
5	$\overline{3}$ 1 $\overline{3}$ $\overline{15}$ 3	55	$\overline{30}$ 1 $\overline{3}$ $\overline{6}$ $\overline{330}$ 6
7	$\overline{4}$ 1 $\overline{2}$ $\overline{4}$ $\overline{28}$ 4	57	$\overline{38}$ 1 $\overline{2}$ $\overline{114}$ 2
9	$\overline{6}$ 1 $\overline{2}$ $\overline{18}$ 2	59	$\overline{36}$ 1 $\overline{2}$ $\overline{12}$ $\overline{18}$ $\overline{236}$ 4 $\overline{531}$ 9
11	$\overline{6}$ 1 $\overline{\overline{3}}$ $\overline{6}$ $\overline{66}$ 6	61	$\overline{40}$ 1 $\overline{2}$ $\overline{40}$ $\overline{244}$ 4 $\overline{488}$ 8 $\overline{610}$ $\overline{10}$
13	$\overline{8}$ 1 $\overline{2}$ $\overline{8}$ $\overline{52}$ 4 $\overline{104}$ 8	63	$\overline{42}$ 1 $\overline{2}$ $\overline{126}$ 2
15	$\overline{10}$ 1 $\overline{2}$ $\overline{30}$ 2	65	$\overline{39}$ 1 $\overline{3}$ $\overline{195}$ 3
17	$\overline{12}$ 1 $\overline{3}$ $\overline{12}$ $\overline{51}$ 3 $\overline{68}$ 4	67	$\overline{40}$ 1 $\overline{2}$ 8 $\overline{20}$ $\overline{335}$ 5 $\overline{736}$ 8
19	$\overline{12}$ 1 $\overline{2}$ $\overline{12}$ $\overline{76}$ 4 $\overline{114}$ 6	69	$\overline{46}$ 1 $\overline{2}$ $\overline{138}$ 2
21	$\overline{14}$ 1 $\overline{2}$ $\overline{42}$ 2	71	$\overline{40}$ 1 $\overline{2}$ $\overline{4}$ $\overline{40}$ $\overline{568}$ 8 $\overline{710}$ $\overline{10}$
23	$\overline{12}$ 1 $\overline{\overline{3}}$ 4 $\overline{276}$ $\overline{12}$	73	$\overline{60}$ 1 $\overline{6}$ $\overline{20}$ $\overline{219}$ 3 $\overline{292}$ 4 $\overline{365}$ 5
25	$\overline{15}$ 1 $\overline{\overline{3}}$ $\overline{75}$ 3	75	$\overline{50}$ 1 $\overline{2}$ $\overline{150}$ 2
27	$\overline{18}$ 1 $\overline{2}$ $\overline{54}$ 2	77	$\overline{44}$ 1 $\overline{2}$ $\overline{4}$ $\overline{308}$ 4
29	$\overline{24}$ 1 $\overline{6}$ $\overline{24}$ $\overline{58}$ 2 $\overline{174}$ 6 $\overline{232}$ 8	79	$\overline{60}$ 1 $\overline{4}$ $\overline{15}$ $\overline{237}$ 3 $\overline{316}$ 4 $\overline{790}$ $\overline{10}$
31	$\overline{20}$ 1 $\overline{2}$ $\overline{20}$ $\overline{124}$ 4 $\overline{155}$ 5	81	$\overline{54}$ 1 $\overline{2}$ $\overline{162}$ 2
33	$\overline{22}$ 1 $\overline{2}$ $\overline{66}$ 2	83	$\overline{60}$ 1 $\overline{3}$ $\overline{20}$ $\overline{332}$ 4 $\overline{415}$ 5 $\overline{498}$ 6
35	$\overline{30}$ 1 $\overline{6}$ $\overline{42}$ $\overline{\overline{3}}$ $\overline{6}$	85	$\overline{51}$ 1 $\overline{\overline{3}}$ $\overline{255}$ 3
37	$\overline{24}$ 1 $\overline{2}$ $\overline{24}$ $\overline{111}$ 3 $\overline{296}$ 8	87	$\overline{58}$ 1 $\overline{2}$ $\overline{174}$ 2
39	$\overline{26}$ 1 $\overline{2}$ $\overline{78}$ 2	89	$\overline{60}$ 1 $\overline{3}$ $\overline{10}$ $\overline{20}$ $\overline{356}$ 4 $\overline{534}$ 6 $\overline{890}$ $\overline{10}$
41	$\overline{24}$ 1 $\overline{3}$ $\overline{24}$ $\overline{246}$ 6 $\overline{328}$ 8	91	$\overline{70}$ 1 $\overline{5}$ $\overline{10}$ $\overline{130}$ $\overline{\overline{3}}$ $\overline{30}$
43	$\overline{42}$ 1 $\overline{42}$ $\overline{86}$ 2 $\overline{129}$ 3 $\overline{301}$ 7	93	$\overline{62}$ 1 $\overline{2}$ $\overline{186}$ 2
45	$\overline{30}$ 1 $\overline{2}$ $\overline{90}$ 2	95	$\overline{60}$ 1 $\overline{2}$ $\overline{12}$ $\overline{380}$ 4 $\overline{570}$ 6
47	$\overline{30}$ 1 $\overline{2}$ $\overline{15}$ $\overline{141}$ 3 $\overline{470}$ $\overline{10}$	97	$\overline{56}$ 1 $\overline{2}$ 8 $\overline{14}$ $\overline{28}$ $\overline{679}$ 7 $\overline{776}$ 8
49	$\overline{28}$ 1 $\overline{2}$ $\overline{4}$ $\overline{196}$ 4	99	$\overline{66}$ 1 $\overline{2}$ $\overline{198}$ 2
51	$\overline{34}$ 1 $\overline{2}$ $\overline{102}$ 2	101	$\overline{101}$ 1 $\overline{202}$ 2 $\overline{303}$ 3 $\overline{606}$ 6

FIGURE 16: Representations $2 \div n$ of the Rhind papyrus

recommendations that characterize certain aspects of the table, but it is not possible to establish the $2 \div n$ table of the Rhind papyrus using his rules only.[10]

Two general observations about the representations found in this table are

(1) the number of unit fractions is kept small within individual representations, and

(2) preference is given to small numerators.

---

10 For a criticism of Gillings' analysis, see Bruckheimer/Salomon, "Some Comments," followed up by Gillings, "Response."

Both of these obviously help to make further calculations as easy as possible. They may also indicate that the $2 \div n$ table is a result of experience and (presumably) trial and error rather than a systematic execution of a set of rules.[11]

The $2 \div n$ table is not the only table concerned with fraction reckoning that is extant. There is also a table of divisions by 10, found in the Rhind papyrus after the $2 \div n$ table, as well as a table of $\frac{2}{3}$ of a fraction (usually listed as number 61 of the Rhind mathematical papyrus).[12] In addition, there is a table for the sums of some Egyptian fractions, which is extant on the mathematical leather roll.

10.3.2 Mathematical Leather Roll

From the techniques of multiplication and division explained earlier, it is obvious that addition of fractions constitutes another technical stumbling block for calculations involving fractions. In executing multiplication or division, potential candidates (to add to yield the factor with which to multiply the other factor or to add to yield the dividend) need to be added up constantly and, if fractions are involved, these need to be summed up—by no means a trivial task because only Egyptian fractions are allowed. Consequently, this is another area of arithmetic in need of tables to facilitate the handling of fractions. Only one of these tables is still extant today, written on a roll of leather (see figure 17). Leather was presumably a prestigious and expensive material to write on. Because leather dries out and becomes extremely brittle, only few leather rolls have survived. They contain different types of texts, including religious (a version of the *Book of the Dead*), literary (a copy of the *Teaching of a Man for His Son*), legal, and administrative texts. The table on the leather roll BM 10250 contains twenty-six sums of up to four fractions, which add up to a single unit fraction. The table is laid out in four columns; however, the first and second columns are simply repeated in the third and fourth columns. The sums and results given in this table are shown in figure 18. It is not clear how the selection of these 26 sums was made. In a few instances (e.g., lines 1 and 2, lines 5 and 7, and lines 23 and 24), when all fractions are even within one line, the next line is obtained by doubling (i.e., halving the denominator). In lines 8 and 9 and lines 25 and 26, the next line is obtained by halving (i.e., doubling the denominator). However, this is not pursued in all possible instances (e.g., line 13: $\overline{12}\ \overline{24} = \overline{8}$ could have been doubled to obtain $\overline{6}\ \overline{12} = \overline{4}$ but was not). Some "doubles" are further away from their possible "origins" (e.g., lines 11 and 20, lines 19 and 25), which makes this explanation less probable.

---

11   For an alternative view, see Abdulaziz, "Egyptian method."
12   On this table, see Neugebauer, *Bruchrechnung*, pp. 36–38.

FIGURE 17: Mathematical leather roll (BM 10250) (© Trustees of the British Museum. All rights reserved.)

1	$\overline{10}\,\overline{40}$	$\overline{8}$
2	$\overline{5}\,\overline{20}$	$\overline{4}$
3	$\overline{4}\,\overline{12}$	$\overline{3}$
4	$\overline{10}\,\overline{10}$	$\overline{5}$
5	$\overline{6}\,\overline{6}$	$\overline{3}$
6	$\overline{6}\,\overline{6}\,\overline{6}$	$\overline{2}$
7	$\overline{3}\,\overline{3}$	$\overline{1\tfrac{1}{2}}$
8	$\overline{25}\,\overline{15}\,\overline{75}\,\overline{200}$	$\overline{8}$
9	$\overline{50}\,\overline{30}\,\overline{150}\,\overline{400}$	$\overline{16}$
10	$\overline{25}\,\overline{50}\,\overline{150}$	$\overline{15}$ (sic text $\overline{6}$)
11	$\overline{9}\,\overline{18}$	$\overline{6}$
12	$\overline{7}\,\overline{14}\,\overline{28}$	$\overline{4}$
13	$\overline{12}\,\overline{24}$	$\overline{8}$
14	$\overline{14}\,\overline{21}\,\overline{42}$	$\overline{7}$
15	$\overline{18}\,\overline{27}\,\overline{54}$	$\overline{9}$
16	$\overline{22}$ (sic text $\overline{12}$) $\overline{33}\,\overline{66}$	$\overline{11}$
17	$\overline{28}\,\overline{49}\,\overline{196}$	$\overline{13}$
18	$\overline{30}\,\overline{45}\,\overline{90}$	$\overline{15}$
19	$\overline{24}\,\overline{48}$	$\overline{16}$
20	$\overline{18}\,\overline{36}$	$\overline{12}$
21	$\overline{21}\,\overline{42}$	$\overline{14}$
22	$\overline{45}\,\overline{90}$	$\overline{30}$
23	$\overline{30}\,\overline{60}$	$\overline{20}$
24	$\overline{15}\,\overline{30}$	$\overline{10}$
25	$\overline{48}\,\overline{96}$	$\overline{32}$
26	$\overline{96}\,\overline{192}$	$\overline{64}$

FIGURE 18: Sums of fractions of the mathematical leather roll

Another observation made by Otto Neugebauer[13] is that many of the sums fulfill the modern equation

$$\tfrac{2}{3}\overline{x} + \tfrac{1}{3}\overline{x} = \overline{x},$$

for example, line 19 (and, likewise, lines 11, 13, 20–26):

---

13   For this and other remarks, see Neugebauer, "Lederrolle." The article has to be read carefully with a copy of the source at hand, as Neugebauer adds sums to the table that are not in the original.

$$\overline{24} + \overline{48} = \overline{16},$$
$$\tfrac{2}{3}\overline{16} + \tfrac{1}{3}\overline{16} = \overline{16}.$$

There are several more modern considerations that can be built on the values we find in the Egyptian text; however, they all fail to explain the selection and order of the table as it is laid out in the source.

10.4 METROLOGICAL TABLES

Another area of mathematics involving tricky computations is the conversion of measures. This is due to subunits of metrological systems, which often involve sets of fractions. For example, the standard measuring unit for lengths was the cubit. However, for smaller lengths, there were palms and digits: 4 digits equal 1 palm and 7 palms equal 1 cubit. Consequently, fractions of a cubit were broken down to palms and digits involving multiplications and divisions by 4 and 7. While arithmetic operations with 4 are fairly straightforward and easy to handle, those with 7 require more attention—as is witnessed by part of an extant table or notes of a calculation in the Ostracon Senmut 153 (New Kingdom)[14] and in number 79 of the Rhind papyrus dealing with multiplications of 7.

Further metrological conversions involve the *ḥqȝ.t* measure and its submultiples $\frac{1}{2}$, $\frac{1}{4}$, $\frac{1}{8}$, $\frac{1}{16}$, $\frac{1}{32}$, and $\frac{1}{64}$. To express amounts of grain smaller than 1 *ḥqȝ.t*, an addition of those submultiples, and—if necessary—the smallest measuring unit 1 *rȝ* ($\overline{320}$ *ḥqȝ.t*) was used. Another capacity-measuring unit was the *hnw*. A table for conversions of these two systems can be found in problem 80 of the Rhind papyrus:

Making in *hnw*

1 *ḥqȝ.t*	10 *hnw*
$\overline{2}$ *ḥqȝ.t*	5 *hnw*
$\overline{4}$ *ḥqȝ.t*	2 $\overline{2}$ *hnw*
$\overline{8}$ *ḥqȝ.t*	1 $\overline{4}$ *hnw*
$\overline{16}$ *ḥqȝ.t*	$\overline{2}\,\overline{8}$ *hnw*
$\overline{32}$ *ḥqȝ.t*	$\overline{4}\,\overline{16}$ *hnw*
$\overline{64}$ *ḥqȝ.t*	$\overline{8}\,\overline{32}$ *hnw*

---

14 Published in Hayes, *Senmut*, p. 29 and plate XXIX; also listed in Fowler, *Mathematics of Plato's Academy*, p. 269.

Fractions of the *ḥq3.t*, if they were not already those submultiples by chance, had to be broken down into them. Obviously, tables were useful in aiding these conversions. In the case of this table, we have a source that documents how the table was constructed. The two wooden boards Cairo CG 25367 and 25368 preserve several groups of calculations like the following:

1	10
[\\] 10	100
[\\] 20	200
[\\] 2	20
1	$\overline{16}\,\overline{32}$ *ḥq3.t* 2 *r3*
\\ 2	$\overline{8}\,\overline{16}$ *ḥq3.t* 4 *r3*
4	$\overline{4}\,\overline{8}\,\overline{64}$ *ḥq3.t* 3 *r3*
\\ 8	$\overline{2}\,\overline{4}\,\overline{32}\,\overline{64}$ *ḥq3.t* 1 *r3*

The first calculation is the division $320 \div 10$, resulting in $10 + 20 + 2$ (32). This can be interpreted as the transformation of $\overline{10}$ *ḥq3.t* into *r3.w*:

$$320\ r3.w = 1\ ḥq3.t,$$
$$\overline{10}\ ḥq3.t = \frac{320}{10}\ r3.w.$$

Note that the techniques chosen (decupling and doubling) produce entries in the first column that can directly be converted into submultiples, for example,

$$10 = \overline{32},$$
$$20 = \overline{16}.$$

The lines that have to be marked for the result are the second, third, and fourth, the second column of which adds up to 320. The respective lines of the first column are 10, 20, and 2. As 10 *r3.w* are $\overline{32}$ *ḥq3.t* and 20 *r3.w* are $\overline{16}$ *ḥq3.t*, the result of the transformation is $\overline{16}$ $\overline{32}$ *ḥq3.t* 2 *r3.w*.

The second calculation is the verification of this result, which is performed by the multiplication $10 \times \overline{16}\,\overline{32}$ *ḥq3.t* 2 *r3.w*. Because $\overline{16}\,\overline{32}$ *ḥq3.t* 2 *r3.w* is supposed to be $\overline{10}$ *ḥq3.t*, the result should be 1 *ḥq3.t*, as it, in fact, is (remember 5 *r3.w* = $\overline{64}$ *ḥq3.t*).

The Cairo wooden boards contain also the respective calculations for $\overline{7}$ *ḥq3.t*, $\overline{11}$ *ḥq3.t*, and $\overline{13}$ *ḥq3.t*.

10.5 SUMMARY

The extant sources allow us to follow Egyptian methods only of multiplication and division. We do not have written information about carrying out addition and subtraction. Multiplication and division are laid out in an identical scheme of two columns. Some of the terminology for multiplication and division also expresses their relation: *wȝḥ-tp m ... r zp ...* "calculate with ... times ..." (multiplication) and *wȝḥ-tp m ... r gm.t ...* "calculate with ... to find ..." (division). Based on these parallels, one could ask if Egyptian mathematics considered multiplication and division as separate arithmetical operations. However, the fact that apart from the expressions mentioned above further verbs are used to designate divisions, e.g., *njs ... ḥnt ...* "call ... before ...", without a match to designate a multiplication, might be taken as a confirmation for the existence of divisions as arithmetical operations in Egyptian mathematics." In the individual performance of a multiplication or division, various techniques could be used.

If fractions are involved, the computation becomes more complicated and often necessitates the use of tables. The conversion of metrological units, as, for example, the *ḥqȝ.t* and its submultiples, also demands the use of tables. A few of these tables are still extant. The most extensive one is the $2 \div n$ table, which is used for the doubling of odd fractions. We do not know how the individual entries of the tables were found; however, in some instances systematic manipulation of previous entries can be observed. Furthermore, the Cairo wooden boards show the construction of part of a metrological table. With the help of few essential tables and a good knowledge of the basic arithmetic operations often needed, a scribe could solve mathematical problems very efficiently, as documents from his daily working life prove.

11.

Mathematics in Practice and Beyond

The mathematical problems and techniques discussed in the previous chapters all come from the small corpus of so-called mathematical texts. These texts were used in the education of scribes, to prepare them for the mathematical tasks they had to cope with in their daily working lives. There are numerous documents on papyri (from the Old Kingdom on) as well as ostraca (mostly from the New Kingdom) that give evidence of the work of these scribes.[1] Written texts include administrative and economic documents and letters but also religious texts as well as literature, some of which, again, comes from an educational context.

The dominant role of mathematics in daily life activities is displayed in other sources as well. Reliefs in tombs are often decorated with scenes from daily life, such as the filling of a granary and the production of bread and beer. Likewise, models left in tombs show the same activities. Scribes are often integral parts of these representations, as, for example, in the model of filling a granary (see figure 19).[2] The workers delivering grain in this model have to pass through a first room in which four scribes record the deliveries, measured by two

---

1   Most notably on papyrus are the Abusir archive and the Gebelein archive (both Old Kingdom); for editions, see Posener-Kriéger, *Archives du temple*, and Posener-Kriéger/Demichelis, *Papiri di Gebelein*. From the Middle Kingdom, the Reisner papyri are an excellent example of documents that show the mathematical practices of the scribes. They were edited (while kept at the Museum of Fine Arts in Boston) in four volumes by William Kelly Simpson (Simpson, *Reisner I–IV*). In 2006, the papyri were moved to the UC Berkeley's Center for the Tebtunis Papyri in the Bancroft Library. Further documents can be found in the Lahun papyri, especially the accounts published in Collier/Quirke, *UCL Lahun Papyri*, vol. 3. A wealth of documents (papyri and ostraca) from the New Kingdom has been discovered at Deir el Medina; see the bibliographies Zonhoven, "Bibliography Deir el Medina," and Haring, "Bibliography Deir el Medina." For an overview of various economic matters for which we have evidence from the individual periods, see also Helck, *Wirtschaftsgeschichte*.

2   The models that were found in the tomb of Meket-Re are among the finest in execution as well as in their state of preservation. Apart from the granary shown in figure 19, they also include a bakery, a slaughterhouse, and a cattle stable. Noteworthy is also the model of a cattle census. A detailed description and photos of these models can be found in Winlock, *Models of Daily Life*.

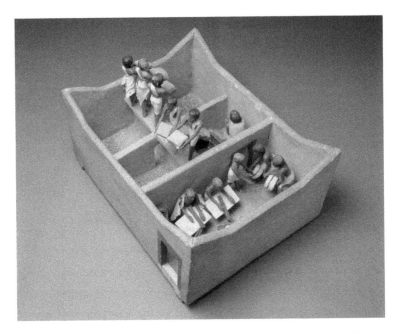

FIGURE 19: Model found in the tomb of Meket-Re (Dynasty 11): Filling of a granary (MMA 20.3.11); Image copyright © The Metropolitan Museum of Art, New York, Rogers Fund and Edward S. Harkness Gift, 1920. Image source: Art Resource, NY.

officials. Then they enter the actual granary to discharge the grain. In models and scenes like this, scribes are depicted in fulfilling their most important duty of accounting for all sorts of goods.[3] For this task, they use the mathematical techniques in which they have been trained. Consequently, the professional output of the scribes, which is extant in administrative and economic documents as well as in letters, includes (results of) mathematical calculations in several instances and thus opens up further possibilities to enlarge our knowledge about ancient Egyptian mathematical practices. However, the mathematics in this type of source is not as easy accessible as that of the mathematical texts. The latter, being used to teach specific mathematical techniques, are very explicit and detailed (and still they sometimes puzzle us!). The problem to be solved is stated explicitly at the beginning. Every step of the solution is indicated by its respective instruction. Often, the solution of the problem is stated explicitly.

---

3   Examples of reliefs and models in which the scribes are shown measuring grain or other products are collected in Pommerening, *Hohlmaße*. For reliefs and paintings, see also Klebs, *Reliefs des Alten Reiches*, Klebs *Reliefs und Malereien des Mittleren Reiches*, and Klebs, *Reliefs und Malereien des Neuen Reiches*.

In some cases, even if instructions are missing, they may be reconstructed by the additionally provided written execution of the respective arithmetic operation.

The situations we encounter in documents resulting from daily mathematical practice are—not surprisingly—different. Instead of problems, instructions, and additional calculations, we are faced only with numbers, and it is up to us to recognize the mathematics behind them. The examples given in the following sections will show that this is not always easy, and it will also show the limitations of information that can be gained. The situation becomes almost hopeless if the source is incomplete, that is, if the papyrus is fragmented. Restoring mathematical operations from data and results alone is difficult enough; if half of these are missing, it simply cannot be done.

However, in some instances information that is not preserved within the corpus of our mathematical texts can be gained from a close study of the evidence extant from a scribe's day-to-day mathematical practice. In order to give a more detailed picture of the situation, the following sections include examples of the respective problems of the mathematical texts—like the Egyptian scribe, we are first taught (some of) the techniques before we encounter the "real" document. This will help the modern reader understand what kind of mathematics is likely to be the origin of the economic or administrative calculation, and it will also accentuate how the two types of sources complement each other.

11.1 DISTRIBUTION OF RATIONS

The distribution of rations constituted a core element of the administration of the Egyptian state.[4] Various amounts, depending on rank, were given to groups of workers, craftsmen, and overseers to ensure that daily life tasks proceeded smoothly. In addition, rations were also distributed within temples (again according to the rank of the beneficiaries) and within individual households. The mathematical side of this practice consists of two kinds of calculations: (1) to determine from an amount that is given for a longer period the respective daily portion, and (2) to distribute an amount of goods among a certain number of beneficiaries, either evenly or according to specific proportions. We can find extant examples for both of these within the mathematical texts as well as within administrative documents. As usual, the mathematical text

---

4   On Middle Kingdom society, its various groups, their relations and other aspects (including the difficulties in interpreting the available source material), see Grajetzki, *Middle Kingdom*, pp. 139–65; Quirke, *Administration*; Quirke, *Titles*, pp. 11–13; on rations (not only Middle Kingdom), see Kemp, *Ancient Egypt*, pp. 117–28.

presents a problem followed by explicit instructions for its solution, whereas the administrative documents may present only the amount to be distributed and the final result.

11.1.1 Ration Problems within the Mathematical Texts

Due to the different types of ration problems, we find several groups of them within the Rhind papyrus. They range from very simple problems that can be solved by executing one division to the most sophisticated problems we find in the extant texts. Surprisingly, despite all the extant papyri, distribution problems occur only in the Rhind papyrus. The Moscow papyrus contains a large number of problems concerned with the production and value of bread and beer but not a single problem about their distribution. The Lahun fragments contain one calculation, which is likely to have been associated with a ration problem. But because only the calculation exists, without any accompanying text, this cannot be verified. The four examples from the Berlin fragments seem not to have contained any ration problems.

The first ration problems of the Rhind papyrus are numbers 1–6, right at the beginning of the problem section. Remember that before these first problems, the $2 \div n$ table and a small table of divisions by 10 were noted. Problems 1–6 introduce the distribution of a number of loaves of bread among 10 men—and are thus an ideal application of the table found immediately before them. The way the problems are noted points to the use of this table. After announcing the problem, that is, stating the numbers of loaves (1, 2, 6, 7, 8, 9) and men (10 in each case), the solution is directly used for verification. This verification is performed as a multiplication of the determined ration by the number of recipients.

PAPYRUS RHIND, NUMBER 6

CALCULATION OF 9 LOAVES OF BREAD for 10 men.
Calculation as it results: Then you multiply $\overline{\overline{3}} \, \overline{5} \, \overline{30}$ times 10.

.	$\overline{\overline{3}} \, \overline{5} \, \overline{30}$
\ 2	$1 \, \overline{\overline{3}} \, \overline{10} \, \overline{30}$
4	$3 \, \overline{2} \, \overline{10}$
\ 8	$7 \, \overline{5}$

Sum: 9 loaves is its result.

The second group of ration problems consists of only two problems, 39 and 40. Again, a given number of loaves of bread are distributed among a given number of persons. This time, however, the distribution is to be unequal. Of the 100 loaves of bread to be distributed among 10 men in number 39, 50 loaves are to be divided among 6 men and the remaining 50 loaves among the other 4 men. The text of the problem does not include any instructions, only the calculations carried out for the solution. This is achieved by two divisions: 50 ÷ 4 and 50 ÷ 6; that is, the unequal distribution is reduced to two equal distributions.

Number 40 is more difficult. Again, 100 loaves of bread are to be distributed. Furthermore, it is specified that the sum of the two lowest rations should be one-seventh of the other three rations:

Papyrus Rhind, Number 40

100 loaves of bread for 5 men, $\overline{7}$ of the 3 upper is for the 2 lower men.

What is the difference?

Calculation as the difference of 5 $\overline{2}$ results:

\.	23	
\.	17 $\overline{2}$	
\.	12	
\.	6 $\overline{2}$	
\.	1,	sum 60.

\.	60
\$\overline{\overline{3}}$	40

You shall multiply 1$\overline{\overline{3}}$ times 23. 38$\overline{3}$ shall result.

.	17 $\overline{2}$	.	29 $\overline{6}$
.	12	.	20
.	6 $\overline{2}$	.	10 $\overline{\overline{3}}$ $\overline{6}$
.	1	.	1 $\overline{\overline{3}}$, sum: 100.

As can be seen from the subsequent calculations, the individual rations should differ by a constant amount 5 $\overline{2}$. Again, we are not given instructions but only written calculations and, in some steps, only the result of calculations performed. The solution apparently starts by assuming a ration of one loaf as the smallest ration. Then the difference between the individual rations is calculated. We are not given any detail of this calculation; only the result is stated as being 5 $\bar{2}$. The individual rations (based on the smallest ration of one loaf and the difference between rations) are calculated and added. As the result (60) is not yet the given number of loaves of bread (100), a factor is determined by which the rations are then multiplied.[5]

The last group of ration problems in the Rhind papyrus can be found in numbers 63–66. Numbers 63–65 are again unequal distributions of loaves of bread or grain among a group of people. Given are the amount of goods to be distributed, the number of recipients, and the ratios of the individual rations. Problem 66 announces the amount of fat for one year; from this, the daily ration must be calculated.

PAPYRUS RHIND, NUMBER 65

METHOD OF CALCULATING 100 LOAVES OF BREAD FOR 10 MEN,
sailor, commander and watchman as double.
Its calculation: You shall add these beneficiaries. 13 shall result.
Divide these 100 loaves of bread by 13. 7 $\bar{\bar{3}}$ $\overline{39}$ shall result.
You shall say: This is the food for these men,
the sailor, the commander and the watchman as double.

.7 $\bar{\bar{3}}$ $\overline{39}$
.7 $\bar{\bar{3}}$ $\overline{39}$
.7 $\bar{\bar{3}}$ $\overline{39}$
.7 $\bar{\bar{3}}$ $\overline{39}$
.7 $\bar{\bar{3}}$ $\overline{39}$
.7 $\bar{\bar{3}}$ $\overline{39}$
.7 $\bar{\bar{3}}$ $\overline{39}$

The sailor 15 $\bar{3}$ $\overline{26}$ $\overline{78}$,
the commander 15 $\bar{3}$ $\overline{26}$ $\overline{78}$,
the watchman 15 $\bar{3}$ $\overline{26}$ $\overline{78}$, sum: 100.

In number 65, 100 loaves of bread are to be distributed among 10 men; however, 3 of these 10 (a sailor, a commander and a watchman) shall receive double rations. The solution simply "counts" these 3 men as 2 persons each (as they receive a double amount). This results

---

5   On this problem, see also Miatello, "A problem of rations."

in a total of 13 portions (7 single and 3 double portions).[6] Thus, the size of an individual portion can be determined by dividing the available amount (100 loaves of bread) by the number of portions (13), which results in 7 $\overline{\overline{3}}$ $\overline{39}$ loaves of bread for each single portion. The individual portions are then listed as they are to be distributed. Therefore, 7 $\overline{\overline{3}}$ $\overline{39}$ is listed 7 times, followed by the double portions (15 $\overline{3}$ $\overline{26}$ $\overline{78}$) for the sailor, the commander, and the watchman. A similar method of establishing rations for employees of different rank can be found in papyrus Berlin 10005, discussed in the next section.

11.1.2 Calculation of Rations in Papyrus Berlin 10005

Papyrus Berlin 10005 was found with other papyri in the pyramid town Lahun, which has also yielded a number of mathematical fragments.[7] The beginning of the extant papyrus includes an offering list, followed by names of priests employed at the funerary temple of Sesostris II. The remainder of the text holds an account of the daily income and expenditure of the temple in bread and two kinds of beer. A translation of this section is given in figure 20.[8]

Part of the income of the temple is used for daily cultic practice. The remainder is then distributed among the personnel of the temple. The account begins with a list of the amounts that are delivered to the temple. These deliveries have two origins: first, there are bread and beer, which are listed as "daily provisions." These constitute the bulk of the daily income (bread: 390 loaves, beer in *sß* vessels: 62, and beer in *ḥpnw* vessels: 172). Secondly, these provisions are supplemented by "what is brought by the temple of Sobek of Crocodilopolis" (bread: 20 loaves, beer in *sß* vessels: 1). The total of these deliveries is indicated. Then, the amounts that are given to the *k3* servants (presumably to carry out the appropriate rituals) are indicated. This is followed by an indication of the remainder. The remainder is then distributed among the temple personnel, who are listed according to their function. Depending on their function, the individuals receive varying amounts; for example, the head of the temple is given 27 $\overline{\overline{3}}$ *ḥpnw* vessels of beer and the pure priest of the king, 11 $\overline{15}$.

A closer look at the individual quantities reveals that the beneficiaries receive amounts according to their position. Thus, an ordinary worker receives much less than the head of the

---

6   For a discussion of further examples of this method from administrative texts, see Mueller, "Wage rates."
7   The hieratic text remains unpublished. Transcription and translation can be found in Borchardt, "Besoldungsverhältnisse," with additions of Gardiner, "Daily income." On the discovery of the papyri, among which Berlin 10005 papyrus was found, see Borchardt, "Papyrusfund."
8   My translation is based on the hieroglyphic text of Borchardt, "Besoldungsverhältnisse," p. 114, and the suggestions/corrections in Gardiner, "Daily income." The translation of the titles is according to Quirke, *Titles*.

Papyrus Berlin 10005

ACCOUNT OF WHAT IS SENT TO THIS TEMPLE			bread	beer $s\underline{t}$ vessels	$\underline{h}pnw$ vessels
Amount of daily provisions		•	390	62	172
What is brought by the temple of Sobek of Crocodilopolis		•	20	1	-
Total			410	63	172
RECKONING AFTER ITS OFFERING (IS TAKEN)					
What is issued to the ka-servants		•	340	28	56 $\bar{2}$
Total			340	28	56 $\bar{2}$
Remainder		•	70	35	115 $\bar{2}$
RECKONING OF THIS REMAINDER			1 $\bar{\bar{3}}$	$\bar{\bar{3}}$ $\bar{6}$	2 $\bar{3}$ $\bar{10}$
Mayor and overseer of the temple	1	10	•	16 $\bar{\bar{3}}$	8 $\bar{3}$	2[7]$\bar{\bar{3}}$
Regulator of a watch in his month	1	3	•	5	2 $\bar{2}$	8 $\bar{5}$ $\bar{10}$
Master lector	1	6	•	10	5	16 $\bar{2}$ $\bar{10}$
Secretary of the temple in his month	1	1 $\bar{3}$	•	2 $\bar{6}$ $\bar{18}$	1 $\bar{9}$	3 $\bar{3}$ $\bar{45}$
Regular lector in his month	1	4	•	6 $\bar{\bar{3}}$	3 $\bar{3}$	11 $\bar{15}$
Embalmer in his month	1	2	•	3 $\bar{3}$	1 $\bar{\bar{3}}$	5 $\bar{2}$ $\bar{30}$
Assistant in his month	1	2	•	3 $\bar{3}$	1 $\bar{\bar{3}}$	5 $\bar{2}$ $\bar{30}$
Libationer in his month	3	2	•	10	5	16 $\bar{2}$ $\bar{10}$
Pure priest of the king in his month	2	2	•	6 $\bar{\bar{3}}$	3 $\bar{\bar{3}}$	11 $\bar{15}$
Medjay (?)	1	1	•	1 $\bar{\bar{3}}$	$\bar{36}$	2 $\bar{3}$ $\bar{10}$
Doorkeeper	4	3	•	2 $\bar{6}$ $\bar{18}$	1 $\bar{9}$	3 $\bar{3}$ $\bar{45}$
Doorkeeper for the night	2	$\bar{3}$	•	1 $\bar{9}$	$\bar{2}$ $\bar{18}$	1 $\bar{23}$ $\bar{90}$
Temple laborer	1	$\bar{3}$	•	$\bar{2}$ $\bar{18}$	$\bar{4}$ $\bar{36}$	$\bar{3}$ $\bar{41}$ $\bar{80}$
Total			70	35	115$\bar{2}$

FIGURE 20: Account of papyrus Berlin 10005

temple. Furthermore, the amounts they receive are fractions or multiples of a basic ration, as has been shown by Michel Guillemot.[9] The basic ration is obtained by dividing the available goods by the sum of the results of the product of the number of persons and their individual multiplication factor. Remember that an identical approach was taken in number 65 of the Rhind papyrus—however, for a less complicated situation.

The available goods are 70 loaves of bread, 35 vessels of a first kind of beer, and 115 $\overline{2}$ vessels of a second type of beer. The number of units (which is obtained by the sum of portions given to the respective positions, each portion being the result of the respective multiplicative factor and the number of people in this position) is 41 $\overline{\overline{3}}$. Therefore, the basic ration amounts to 1 $\overline{\overline{3}}$ $\overline{75}$ (that is, 70 divided by 41 $\overline{\overline{3}}$) loaves of bread, $\overline{\overline{3}}$ $\overline{6}$ $\overline{150}$ (that is, 35 divided by 41 $\overline{\overline{3}}$) vessels of the first type of beer, and 2 $\overline{\overline{3}}$ $\overline{10}$ $\overline{250}$ $\overline{750}$ (that is, 115 $\overline{2}$ divided by 41 $\overline{\overline{3}}$) vessels of the second type of beer.

Note that the amount of the first type of beer is half of the amount of bread (since there is double the number of loaves of bread compared to the number of vessels of beer). To calculate the actual rations that are noted in the papyrus, the rounded figures 1 $\overline{\overline{3}}$ and 2 $\overline{\overline{3}}$ $\overline{10}$ are used. Papyrus Berlin 10005 notes the number of persons belonging to one profession (and, hence, the same ration level), the factor by which to multiply the basic ration, and the totals of given goods. The individual factors and the respective amounts of beer are shown in table 6.

11.1.3 Further Evidence

Several other sources are extant that detail rations in various contexts—for example, the administration of a temple, daily life at a palace, a building project, a private household, and a military excursion. Papyrus Boulaq 18 presents an account of the daily income and expenditure at the palace in Thebes over 12 days.[10] Papyrus Reisner I, section O, contains a list of 20 persons and the quantities of bread they received, laid out in a tabular format.[11] This ration list is part of the documentation of a building project. Papyrus MMA 22.3.517 (Letter 2), a letter written by the priest Hekanakhte during his absence from home, lists the distribution of a monthly amount of grain among several members of his household.[12] Papyrus

9 Guillemot, "Notations."
10 For a facsimile of the hieratic text, see Mariette, *Papyrus Boulaq*, pp. 14ff. For a transcription and translation (German) see Scharff, "Rechnungsbuch." For an English translation and a discussion of the text see Quirke, *Administration*, pp. 9–50.
11 Hieratic text, translation, and commentary can be found in Simpson, *Reisner I*, plates 21 and 21a and p. 131. See also Imhausen, "Egyptian mathematics," pp. 40–44.
12 Hieratic text and translation can be found in Allen, *Heqanakht*, pl. 10–11 (photo), pl. 30-33 (facsimile and hieroglyphic transcription) and pp. 16–17 (translation).

TABLE 6: Rations in papyrus Berlin 10005

	number of persons	factor	beer (*ḥpnw*) allocated to "group"
temple laborer	1	$\overline{3}$	$\overline{\overline{3}}\,\overline{4}\,\overline{180}$
doorkeeper for the night	2	$\overline{3}$	$1\,\overline{2}\,\overline{3}\,\overline{90}$
unit		1	$2\,\overline{\overline{3}}\,\overline{10}$
secretary of the temple	1	$1\,\overline{3}$	$3\,\overline{\overline{3}}\,\overline{45}$
embalmer	1	2	$5\,\overline{2}\,\overline{30}$
regulator of a watch	1	3	$8\,\overline{5}\,\overline{10}$
regular lector	1	4	$11\,\overline{15}$
master lector	1	6	$16\,\overline{2}\,\overline{10}$
mayor and overseer of the temple	1	10	$27\,\overline{\overline{3}}$

Harhotpe IX is a fragmentary list of quantities of dates and two kinds of grain given to a group of soldiers.[13]

Despite the variety of contexts, there are several common features among them. The records are all written on papyrus and clearly come from an economic/administrative background. All use the Egyptian word ⸢q.w for the goods that are distributed.[14] They display the framework that was used to distribute amounts of food to a group of people, which may be hierarchically structured. The structure of the individual group is mirrored in the amounts that individual positions receive. This concept is also part of the mathematical training, as problems 40 or 65 of the Rhind mathematical papyrus illustrate.

The information about the absolute numerical values of food supply that can be drawn from this evidence is limited.[15] Often it is not obvious if the beneficiaries of those documents—if they are explicitly mentioned at all—have a further income. The exact period for which the noted rations are intended to last is not always mentioned. Furthermore, it is not clear if the specified amount of food is meant only for the listed beneficiary or for his family as well. These difficulties arise, as so often, from the source situation. Only isolated pieces of text,

---

13   Hieratic text and translation and commentary can be found in James, *Hekanakhte*, pl. 16-17 and
       pp. 71–74.
14   For a discussion of the term ⸢q.w, see Allen, *Heqanakht*, pp. 145–46.
15   For a discussion, see Imhausen, "Calculating the daily bread."

without their original contexts, are extant. If only we had the complete documentation of the grain that Hekanakhte administered.

However, even with the limited evidence at hand, it is possible to link a mathematical technique found in the Rhind papyrus to the mathematics used to establish actual rations in practice in papyrus Berlin 10005. Moreover, the various contexts of existing ration texts from the Middle Kingdom document the use of mathematics in these contexts.

11.2 ARCHITECTURAL CALCULATIONS

Among the extant artifacts from ancient Egypt, the remains of temples and tombs are probably the most impressive. In order to accomplish this level of architectural achievements, numerous workers were needed, as well as efficient mathematical techniques in order to carry out the planning and organization of such projects. The extant evidence for architectural calculations is rather limited. Some information can be gained from inscriptions in quarries detailing workers involved and the amount of stones removed, as, for example, those from the Wadi Hammamat.[16] Within the mathematical papyri, only six problems related to architecture can be found (Rhind papyrus, numbers 56–60 and Moscow papyrus, number 14).[17] Evidence of architects from the Middle Kingdom is rather sparse. Few instruments are extant, and only one papyrus, papyrus Reisner I, contains the documentation of a temple-building project. While there is no such documentary evidence for any of the extant temples, the temple whose construction is described in papyrus Reisner I has not yet been discovered.[18]

11.2.1 Architectural Problems within the Mathematical Texts

Apart from number 14 of the Moscow papyrus, the calculation of the volume of a truncated pyramid, all extant examples are found in the Rhind papyrus.[19] They are grouped together as numbers 56–60. All teach the relation between base, height, and inclination of a structure with slanted sides. In order to express the inclination of a surface, the Egyptians used the concept of the *sqd* (see figure 21). The *sqd* " . . . was the horizontal displacement of the sloping face

16   See Goyon, *Wadi Hammamat*, no. 61, pp. 17–20 and pp. 81–85, as well as Couyat/Montet, *Ouâdi Hammâmât*, nos. 123 and 192.

17   Another possible architectural element is featured in number 10 of the Moscow papyrus. But as this is the calculation of the surface of an object (most likely a cylinder), it is rather difficult to put this problem into a practical context. Recent discussions of this problem can be found in Hoffmann, "Aufgabe 10"; Imhausen, *Algorithmen*, pp. 76–77, and Miatello, "Moscow mathematical papyrus and a tomb model."

18   On the mathematics and architecture of Papyrus Reisner I see also Rossi/Imhausen, "Papyrus Reisner I."

19   For a discussion of the algorithms of these problems, see Imhausen, *Algorithmen*, pp. 162–68.

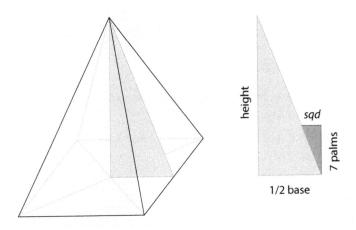

FIGURE 21: Description of sloping surfaces with the *sqd*

for a vertical drop of one cubit, that is, the number of cubits, palms, and fingers by which the sloping side had 'moved' from the vertical at the height of one cubit."[20]

The *sqd* is always indicated in palms and, if necessary, digits. Evidence for this way of handling sloping surfaces can already be found in the Old Kingdom.[21] Numbers 56–59 of the Rhind papyrus address the calculation of pyramids (*mr*):

PAPYRUS RHIND, NUMBER 56

[1] METHOD OF CALCULATING a pyramid (of) 360 as base and 250 as its height.

[2] Let me know its *sqd*.

You shall calculate half of 360. 180 shall result.

You shall divide [3] 180 by 250. $\overline{2}$ $\overline{5}$ $\overline{50}$ of a cubit shall result.

1 cubit is 7 palms.

You shall multiply with 7.

[4] .   7

[5] $\overline{2}$ 3 $\overline{2}$

[6] $\overline{5}$ 1 $\overline{3}$ $\overline{15}$

20   Rossi, *Architecture and Mathematics*, p. 185.
21   Around the corners of Mastaba 17 at Meidum, the excavator William M. F. Petrie found a series of diagrams that indicated the slope of the sides of the mastaba. Four L-shaped mud-brick walls had been built beneath the ground level around the four corners of the mastaba. On these walls, horizontal lines spaced 1 cubit (or 7 palms) apart were indicated, as well as the slope of the mastaba. For a depiction and detailed explanation see Arnold, *Building*, p. 12, and Rossi, *Architecture and Mathematics*, pp. 188–92.

^{7} $\overline{50}$ $\overline{10}\,\overline{25}$

^{8} Its *sqd* is ^{9} 5 $\overline{25}$ palms.

Papyrus Rhind, Number 57

^{1} A pyramid: 140 is the base and 5 palms 1 digit is its *sqd*.

What is its respective height?

^{2} You shall divide 1 cubit by the *sqd* times 2. 10 $\overline{2}$ shall result.

You shall divide ^{3} 7 by 10 $\overline{2}$.

Look, it is 1 cubit.

Calculate with 10 $\overline{2}$.

$\overline{\overline{3}}$ of 10 $\overline{2}$ is 7.

^{4} You shall calculate with 140. This is the base.

Calculate $\overline{\overline{3}}$ of 140 as 93 $\overline{3}$.

^{5} Look, it is its respective height.

Numbers 56 and 57 of the Rhind mathematical papyrus show two variations of the *sqd* problems. In number 56, the base and height of the pyramid are indicated; the resulting inclination of the sides (*sqd*) is determined. In number 57, base and *sqd* are indicated, and the height of the pyramid with these specifications is determined. The procedures of both problems include the conversion of cubits into palms and digits, or vice versa. This conversion is explicitly indicated as such within the text of both problems (number 56: 1 cubit is 7 palms; number 57: Look, it is 1 cubit).

After each problem, a display drawing illustrates the object that is calculated. The pyramids of numbers 56–59 are each depicted in their characteristic triangular side view (see figure 22). The top of the pyramid is indicated as a solid colored triangle. All pyramids seem to be placed upon a base that extends beyond their sides. Despite these detailed illustrations, there is no reference in the text concerning specific details of either the top or base construction of the pyramids.[22]

The problems related to pyramids are followed by a problem that is also concerned with the relation between base, height and the inclination of the slanted sides:

---

22 Note that the basic style of construction of a pyramid changes from the Old to the Middle Kingdom. From Sesostris I on, the pyramids have an "internal skeleton of limestone walls forming compartments filled with roughly shaped stones" (Lehner, *Complete Pyramids*, p. 170). For a general overview of Egyptian pyramids see Lehner, *Complete Pyramids*, and Verner, *Pyramids*.

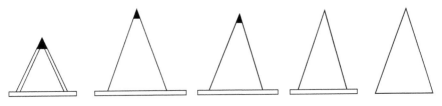

FIGURE 22: Drawings of pyramids from problems 56–60 of the Rhind papyrus

PAPYRUS RHIND, NUMBER 60

[1] A PILLAR of 15 cubits as its base, 30 as its height to the upper side.

[2] Let me know its *sqd*.

Calculate with 15. Its half is 7 $\overline{2}$.

Multiply [3] 7 $\overline{2}$ by 4 to obtain 30.

Its *stwtj* shall result as 4. This is [4] its respective *sqd*.

[5] Calculation:

[6]	.	15
[7]	\\$\overline{2}$	7 $\overline{2}$
[8]	.	7 $\overline{2}$
[9]	2	15
[10]	\4	30

The object of this problem is called *jwn* in the Egyptian text, here translated as "pillar."[23] The procedure begins like those of the previous problems, with the calculation of half of the base. To calculate the *sqd*, as requested in the problem, this (half of the base) should have been divided by the height. Instead, however, the height is divided by the result of the first step (half of the base), which yields the inverse of the *sqd*, which is designated as *stwtj*.[24] The text of the problem then states that this is "its respective *sqd*." The reason for the calculation of the inverse of the *sqd* may originate from the type of object and the numerical values involved. The sides of the object are much steeper than those of the previous pyramids. Their actual *sqd* is 1$\overline{2}$ $\overline{4}$. If this were to be used in practice, a minimal deviation would result in a relatively large

---

23   Note that the drawing for this problem renders it as if it were a pyramid. For a discussion of the term *jwn* ("column"), see Spencer, *Temple*, pp. 231–35. Further evidence for *jwn* designating an object with a square base can be found in Anastasi I papyrus, where it designates the shaft of an obelisk. Another discussion of the problem is Miatello, "Problem 60 of the Rhind mathematical papyrus."

24   The meaning of *stwtj* is not straightforward, and hence it is best left untranslated. For a preliminary discussion of the term, see Simpson, *Reisner I*, pp. 78–79.

error. Hence, the use of the inverse of the *sqd* in problem 60 of papyrus Rhind may result from practical implications.

11.2.2 Mathematics of Papyrus Reisner I

Examples of other evidence for the time of the Middle Kingdom are the so-called Reisner papyri, named after the Egyptologist who found them, George Reisner. During his first excavation of Naga ed Deir in 1904, Reisner discovered four rolls of papyri, which were lying on top of one of the coffins. There is no indication why these papyri were left there. They date from the Middle Kingdom and are thus contemporary with the majority of the hieratic mathematical texts.[25] They contain the documentation of a building project in the time of King Sesostris I as well as accounts of a shipyard.[26]

The Reisner papyri complement the mathematical texts in several aspects. They document the administration of work; for example, they demonstrate how a workload was mathematized and the number of men that were determined to be needed to complete a task in a certain time.[27] Furthermore, the calculations noted in the Reisner papyri also give us further insights into Egyptian metrology. The mathematical texts provide us with information about capacity units used for measuring grain and liquids. However, there are no units indicated for the measurement of volumes, which are not grain products or liquids. These are provided by the Reisner papyri.

Within papyrus Reisner I, several sections table calculations of the volume of material that is moved in one phase of the project. The text is divided into several parts, each of which has a header that mentions a date and specifies the step of the project. Below these headers, we find tabular arrangements of numbers, which follow the same structure in each section.

The first column indicates the length, the second column, the width, the third, the depth. These three dimensions characterize the size of a block of material that is moved. The fourth column indicates the number of blocks of this size, and the last column lists the total volume of blocks of the respective size, which is calculated as the product of the number of blocks and the volume of one block, which can be determined from the three given dimensions given in the first three columns.

---

25   Until 2006, the Reisner papyri were kept at the Museum of Fine Arts in Boston. They are now held at UC Berkeley's Center for the Tebtunis papyri in the Bancroft Library.

26   They have been edited by William Kelly Simpson in four volumes; see Simpson, *Reisner I, Reisner II, Reisner III,* and *Reisner IV.* The most interesting of these from the point of view of a historian of mathematics is doubtless papyrus Reisner I, edited in the first volume. For a discussion of some aspects see Rossi/Imhausen, "Papyrus Reisner I."

27   On the organization of work at various times during the ancient Egyptian history see Menu, *Organisation du travail.*

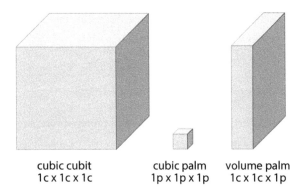

cubic cubit   cubic palm   volume palm
1c x 1c x 1c   1p x 1p x 1p   1c x 1c x 1p

FIGURE 23: Volume cubit (cubic cubit), cubic palm, and Egyptian volume palm

The measurements given in the first three columns are sometimes cubits alone, and sometimes they consist of cubits and the smaller units, palms and digits. The entries in the last column, which lists the volumes, are written in the same way the linear measurements are indicated, as cubits, palms, and digits, which obviously have to be understood as volume measurements: that is, volume cubits, volume palms, and volume digits. Therefore, this table is the evidence of the use of a cubic cubit, which is not attested to from the volume calculations of the mathematical texts. The cubits in this column, as in the linear measurements, are always denoted by their numerical value alone; only palms and digits are indicated as such. From the numerical values involved, it is possible to deduce in which way these units of volume were defined. While the volume cubit is—as we would expect—a cube of length, width, and height 1 cubit, the volume palm is a cuboid that is 1 cubit by 1 cubit by 1 palm, and likewise, the volume digit is a cuboid of 1 cubit by 1 cubit by 1 digit (cf. figure 23).

11.3 LAND MEASUREMENT

The surveying of land was put forward as the reason for the development of mathematics in Egypt by its first historiographer Herodotus:

> This king also (they said) divided the country among all the Egyptians by giving each an equal parcel of land, and made this his source of revenue, assessing the payment of a yearly tax. And any man who was robbed by the river of part of his land could come to Sesostris and declare what had happened; then the king would send men to look into it and calculate the part by which the land was diminished, so that

thereafter it should pay in proportion to the tax originally imposed. From this, in my opinion, the Greeks learned the art of measuring land.[28]

11.3.1 Area Calculations in the Mathematical Texts

In the Rhind papyrus, numbers 48–53 determine areas of various shapes, and numbers 54 and 55 divide a given area into a number of smaller areas. The technical term for surface, or area, used in numbers 49–52 is $3h.t$, the original meaning of which is "field." Numbers 48 and 53 show only a drawing and calculations but no text. Still, their placement with numbers 49–52 indicates that they were conceived to belong to the same group of problems. Despite the use of the term $3h.t$ for area, it is not possible to classify numbers 48–53 straightforwardly as the calculation of areas of fields. The structures that are the subject of number 48 are a circle and a square. While a square field is perfectly reasonable, a circular one is rather hard to imagine.[29] While the technical term $3h.t$ for area suggests an origin of calculations of areas with the determination of the areas of fields, as proposed by Herodotus, the evidence of the hieratic mathematical papyri demonstrates that it had since gone beyond practical calculations (of fields) only. The use of the same technical term and the grouping of the area calculations as one block within the Rhind papyrus demonstrate that the geometric concept of the determination of an area was the decisive factor for classifying a problem within this group and not its possible practical context.

However, the calculation of areas of fields remained an important application of these calculations, as evidenced by sources from the Middle Kingdom until the Greco-Roman Periods. The following examples are calculations of various areas found in the hieratic mathematical papyri. Problem 50 of the Rhind papyrus is the calculation of the area of a circle.

Papyrus Rhind, Number 50

[1] Method of calculating a circular area of 9 ht.

[2] What is its value/amount as an area.

You shall subtract its $\overline{9}$ as 1, [3] the remainder is 8.

You shall multiply 8 by 8. [4] 64 shall result.

Its value/amount as an area is 64.

[5] Calculating according to the procedure:

---

28 Herodotus, Histories II, 109, translation taken from Godley, *Herodotus*. For a discussion of this passage and others of a similar kind, see Fowler, *Mathematics of Plato's Academy*, pp. 279–80, and Cuomo, *Ancient Mathematics*, p. 6. For an introduction and commentary of Herodotus' account of Egypt in general, see Lloyd, *Herodotus Book II*.

29 Because of this, I included the area calculations within the problems training basic mathematical techniques in Imhausen, *Algorithmen*.

[6]	.	9
[7]	$\overline{9}$	1

[8] Subtracting from it, remainder: 8.

[9]	.	8
[10]	2	16
[11]	4	32
[12]	8	64

Its value/amount as area: 64.

Note that this procedure does not involve any notion of π. In order to judge the exactness of the Egyptian procedure, it can be transformed into a modern equation (A = area, d = diameter), which yields the same result as the Egyptian procedure:

$$A_{\text{circle, Egyptian}} = (d - \tfrac{1}{9}d)^2.$$

For a comparison with our modern calculation of the area of a circle, this equation can be transformed to match the modern formula,

$$A_{\text{circle, modern}} = \pi \left(\tfrac{d}{2}\right)^2 = \tfrac{\pi}{4}d^2,$$

as

$$A_{\text{circle, Egyptian}} = \left(\tfrac{8}{9}d\right)^2 = \tfrac{64}{81}d^2.$$

The comparison of the two equations shows that where the modern formula uses $\tfrac{\pi}{4}$, the equation originating from the Egyptian procedure has $\tfrac{64}{81}$. Thus, the π of the modern formula is equivalent in the modern rewriting of the Egyptian procedure as $\tfrac{256}{81}$, that is, approximately 3.16.

Note, however, that these considerations are useful only to estimate how accurate the Egyptian procedure was compared to our modern formula. The procedure given in the Egyptian sources indicates that the relation of diameter and circumference did not play any role in establishing the area of a circle. The extant types of problems with circles further corroborate this. Contrary to the pyramid problems, where individual problems calculate one of three parameters that are linked (base of a pyramid, height of a pyramid, inclination of a pyramid), the Egyptian procedure calculates the area of a circle only from a given diameter. The circumference is not even mentioned in any of the hieratic mathematical problems.

Rewriting the procedure in form of its symbolic algorithm yields the following:

	9			D_1
(1)	$\bar{9} \times 9 = 1$		(1)	$\bar{9} \times D_1$
(2)	$9 - 1 = 8$		(2)	$D_1 - (1)$
(3)	$8^2 = 64$		(3)	$(2)^2$

The only constant used in this procedure is $\bar{9}$, which made 9 a preferred numerical value for the diameter of a circle (see also numbers 41 and 48 of the Rhind papyrus). The constant $\bar{9}$ is also used in the much-disputed number 10 of the Moscow papyrus:

Papyrus Moscow, Number 10

[1] Method of calculating a nb.t.

[2] If you are told:

a nb.t (of 4 $\bar{2}$) as (its) base,

[3] by of 4 $\bar{2}$ as (its) side.

Oh, [4] let me know its area.

You shall calculate [5] $\bar{9}$ of 9,

because, concerning a nb.t: [6] It is half of an (. . .).

1 shall result.

[7] You shall calculate the remainder as 8.

[8] You shall calculate $\bar{9}$ of 8. [9] $\bar{3}\ \bar{6}\ \overline{18}$ shall result.

You shall calculate [10] the remainder of these 8 after these [11] $\bar{3}\ \bar{6}\ \overline{18}$.

7 $\bar{9}$ shall result.

[12] You shall calculate 7 $\bar{9}$ times 4 $\bar{2}$. [13] 32 shall result.

Look, it is its area.

[14] What has been found by you is correct.

The object of which the surface is calculated in this problem is designated by the Egyptian word nb.t. Elsewhere, nb.t is attested to as "basket," and it can be assumed that the mathematical shape is at least related to that of a basket. Based on the following text, the shapes of a hemisphere, a half cylinder, a half circle, and an elongated dome have been suggested as that of a nb.t.[30] The procedure is similar to that of the calculation of the area of a circle, involving

---

30   Hemisphere: Struve, *Mathematischer Papyrus Moskau*, pp. 157–69; Gillings, "Hemisphere"; Fletcher, "Hemisphere"; Couchoud, "Surfaces rondes"; half cylinder: Peet, "Egyptian geometry"; Hoffmann,

the calculation of $\overline{9}$ of one of the data and its subtraction.[31] However, squaring the result of this step does not follow, as would be expected in the case of a circle; instead $\overline{9}$ of the remainder is again calculated.

The similarity to the procedure of calculating the area of a circle points to a curved object, which is true for all of the previously listed possibilities. The further identification of the object is hindered by an obvious flaw of the source text: although two technical terms (referring to two dimensions of the object) are explicitly indicated, only one numerical value is specified. The following procedure seems to imply that the two dimensions were equal. The various possibilities have been discussed in detail by Friedhelm Hoffmann, who endorsed the half cylinder that was first suggested by Thomas Eric Peet.[32] Another oddity of this problem is its lack of a practical background. Mostly, the Egyptian mathematical problems without a practical background are those that serve the training of basic calculation techniques, such as, for example, fraction reckoning or the calculation of an unknown from a given mathematical relation of that unknown ($ʿḥʿ$ problems). However, further speculations about this problem seem futile as long as the object that is calculated cannot be ascertained.[33]

The following calculation of the area of a triangle may well have been used in an actual practical context. It is extant twice, once in the Rhind papyrus (number 51) and, somewhat damaged, in the Moscow papyrus (number 4).

PAPYRUS RHIND, NUMBER 51

[1] METHOD OF CALCULATING A TRIANGLE AS AREA

[2] If you are told:

a triangle

of 10 $ḫt$ as [3] height, 4 $ḫt$ as its base.

[4] What is its area?

Calculating according to the procedure:

[5] You shall calculate half of 4 as 2,

[6] to cause it to be half a rectangle.

[7] You shall multiply 10 [8] times 2.

It is its area.

"Aufgabe 10"; Miatello, "Moscow mathematical papyrus and a tomb model"; half circle: Peet, "Egyptian geometry"; elongated dome: Neugebauer, "Geometrie", pp. 427–28.

31   For a detailed discussion see Imhausen, *Algorithmen*, pp. 76–77.

32   Hoffmann, "Aufgabe 10."

33   Miatello, "Moscow mathematical papyrus and a tomb model," also attempts an identification, which would link it to the area of volume measures.

Despite the damaged state of number 4 of the Moscow papyrus, it can be identified as the same problem, even with identical numerical values:

PAPYRUS MOSCOW, NUMBER 4

[1] [METHOD OF] CALCULATING A TRIANGLE

[2] If you are told:

a triangle of height 10 and

[3] 4 as its base.

You shall let me know [4] [...]

[...] this 2.

[5] [...] times 2.

[6] It is [...].

The area of the triangle is determined as the product of half of the base and the height. The text of the Rhind papyrus even indicates a motivation behind this procedure: The area of a triangle is perceived as half the area of a rectangle with the base and height of the triangle as its sides.

Other area problems of the Moscow papyrus involve the relation of the sides of a rectangle (Moscow papyrus, number 6) and rectangular triangles (Moscow papyrus, numbers 7 and 17):

PAPYRUS MOSCOW, NUMBER 17

[1] Method of calculating a triangle.

[2] If you are told:

a triangle of 20 as its area.

[3] Concerning that which you put as the length,

you have put $\overline{3}\ \overline{15}$ and it is the width.

[4] You shall double 20.

40 shall result.

[5] You shall divide 1 by $\overline{3}\ \overline{15}$.

2 $\overline{2}$ times shall result.

[6] You shall calculate 40 times 2 $\overline{2}$.

100 shall result.

You shall calculate its square root.

[7] 10 shall result.

Behold, it is 10 as the length.

You shall calculate $\overline{3}\,\overline{15}$ [8] of 10.

4 shall result.

Behold it is 4 as the width.

What has been found by you is correct.

Rewriting the procedure in form of its symbolic algorithm yields:

20		D_1	
$\overline{3}\,\overline{15}$		D_2	
(1)	$2 \times 20 = 40$	(1)	$2 \times D_1$
(2)	$1 \div \overline{3}\,\overline{15} = 2\,\overline{2}$	(2)	$1 \div D_2$
(3)	$40 \times 2\,\overline{2} = 100$	(3)	$(1) \times (2)$
(4)	$\sqrt{100} = 10$	(4)	$\sqrt{(3)}$
(5)	$\overline{3}\,\overline{15} \times 10 = 4$	(5)	$D_2 \times (4)$

The procedure reveals that the triangle of this problem was a rectangular triangle, and the ratio was that of the sides including the right angle. The procedure begins by calculating the area of the rectangle obtained as double the area of the triangle (1), that is, a rectangle with the legs of the right-angled triangle as sides. This is followed by calculating the inverse of the ratio of the sides (2), which is then multiplied by the area of the rectangle (3), resulting in the area of a square, with the side equal to the longer side of the rectangle. The extraction of the square root yields the length of this side (4), and multiplication of this side with the ratio then yields the length of the remaining side. Although the triangle is rectangular and the solution makes use of this property, there is no evidence for the use of the rule of Pythagoras.[34] The same is true for the other instance of a rectangular triangle (number 7 of the Moscow papyrus).

11.3.2 Evidence of Land Surveying

In the New Kingdom, the measuring of fields is depicted several times in Theban tombs (TT).[35] From the Middle Kingdom, papyrus Harageh 3, one page of a journal accounting for the measuring of fields over several days, is extant.

---

34   I use the distinction between the Pythagorean theorem and the Pythagorean rule as established by Jens Høyrup in "Pythagorean rule and theorem."
35   See, for example, TT38, TT57, TT69, TT75, TT86, and TT297, some of which are discussed in Berger, "Land measurement."

Papyrus Harageh 3 is today kept in the Petrie Museum as papyrus UC 32775 and dates to the late Twelfth Dynasty:

PAPYRUS HARAGEH 3[36]

[...]

[The secretary] of fields		Senebtj[fj]
		Seneb
Commissioner of the overseer of the estate [...]		Hori
Man who stretches the cords		Satephu
Man who takes the cords		Ibj
Year 2, SECOND MONTH OF INUNDATION,	DAY 15	Spent in
	DAY 16	an enquiry
	DAY 17	in the bureau of
	DAY 18	fields of the
	DAY 19	Southern District
Year 2, SECOND MONTH OF INUNDATION,	DAY 20	

SPENT recording for it the assessment of income due, in the bureau of fields of the Northern District.

ASSEMBLY BEFORE THE NATIONAL OVERSEER OF FIELDS REDIENPTAH, NORTHERN DISTRICT.

LIST BY NAMES OF THE SECRETARIES OF FIELDS WHO HAVE COME FOR THE REGISTRATION TODAY. REGISTERING BY THE SCRIBE OF THE CADASTER AND KEEPER OF REGULATIONS PAENTIENI:

Secretary of the fields		Senebtjf[j]
		Seneb
Commissioner of the overseer of the estate [...]		Hori
Man who takes the cords		Ibj
Man who stretches the cords		Satephu
Year 2, SECOND MONTH OF INUNDATION,	DAY 21	Spent
	DAY 22	in
	DAY 23	[...] income due, in the bureau
		of the fields [...]

---

36 The text was published in Smither, "Tax-assessor's journal." The translation given here is based on that given by Smither and that of Quirke, *Administration*, pp. 174–75; note the emendations given in Quirke, *Administration*, p. 186, note 65. The titles are (when possible), according to Quirke, *Titles*. For a discussion of this text, see also Quirke, *Administration*, pp. 174–76.

It is assumed that this registration occurred within the context of grain taxation.[37] The officials involved consisted of a team of various persons with individual designations, such as "scribe of the cadaster and keeper of regulations" (*zš n tm3 jry ḥp*), two "scribes of the fields" (*zš 3ḥ.wt*), a "holder of the cord" (*šzp nwḥ*) and a "stretcher of the cord" (*dwn nwḥ*). There is no information about the actual measuring procedure—a substantial part of the process seems to have taken place in the "bureau of the fields." The number of people involved may have served to control the registration, or, as Stephen Quirke described it, "administrative procedure requires one official to oversee the act, while a second performs the task of recording it."[38]

11.4 SUMMARY

The practical problems constitute more than half of the extant mathematical problem texts. Furthermore, most of the more advanced problems are those with a real or pseudo practical background. This is not surprising if the evidence for mathematics beyond the mathematical texts is taken into account. Scribes were confronted with mathematical problems throughout their work lives, as is depicted in reliefs and tomb models. There can be little doubt that the extant mathematical texts were aimed at providing the scribes with the necessary mathematical techniques for exactly this type of work—hence the dominance of practical problems.

The problem group with the most examples within the mathematical texts is that of the bread-and-beer problems.[39] They indicate the mathematical framework that was used to assess the production of bread and beer from grain and are part of the mathematical problems related to the management of grain production.[40] The bread-and-beer problems were not included in the preceding examples because there is no extant match in the administrative documents from the time of the Middle Kingdom.[41] No account from a group of bakers or brewers is extant from this time; therefore, the mathematical problems can be linked only to the tomb depictions showing baking and brewing. In some of these depictions, grain is taken by a scribe

37 Smither, "Tax-assessor's journal," p. 76.
38 Quirke, *Administration*, p. 175.
39 For a detailed discussion of those and their background, see Imhausen, *Algorithmen*, pp. 111–38 and Imhausen, "Calculating the daily bread."
40 For an overview of grain production, see Murray, "Cereal production," or Faltings, *Lebensmittelproduktion*. For an overview of the accounting of the grain production in its various stages (up to the delivery of its products in the form of bread and beer), see Imhausen, "Mathematisierung von Getreide."
41 From the later New Kingdom (from the time of Seti I), accounts of bakers are extant (for an edition, see Spiegelberg, *Rechnungen*). However, they seem to use a different conceptual framework from that, which is used within the mathematical texts. The loaves of bread are characterized by their weight (indicated in *dbn*). The technical term *psw*, which is the characteristic element of the hieratic bread-and-beer problems, is still attested within these texts, but according to Anthony Spalinger, the term seems to designate a loss of substance during baking (see Spalinger, "Baking during the reign of Seti I").

TABLE 7: Overview of practical problems and their terminology

topic	Egyptian term	pRhind	pMoscow	pLahun	pBerlin
area	*ꜣḥ.t*	8	5	1	-
rations	*ꜥ.w/pzš*	15	-	(1)	-
bread-and-beer	*psw*	10	11	-	-
granaries	*šꜥ*	6	-	1	-
work produce	*bꜣk.w*	2	2	1	-
value of a bag of metals	*qrf.t*	1	-	-	-
pyramid properties	*mr/jwn*	6	-	-	-

from a granary and given to a worker, while another scribe is noting the amount that has been handed over.[42] This is followed by depictions showing various stages of baking and brewing, until finally the products are brought for measuring. Thus, these depictions indicate the framework of the problems found in the mathematical texts: a specified amount of grain must enable the scribe to determine the amounts of products that can be expected.

Within the problem text, a specific terminology is used, consisting not only of technical terms for various arithmetic operations, but also of terms that indicate the affiliation of a problem to a specific group. This key term is often employed in the title of a problem and thereby refers the junior scribe to a specific area of his mathematical training. For an overview of these terms and their associated problems, see table 7.

The realm of practical problems also enables a historian to use supplementary evidence to complement the information from the mathematical texts. This has been especially effective in the case of the architectural problems. Only half a dozen architectural problems are extant within the mathematical texts—and with five of them being variations of one problem, this constitutes a rather meager yield if one considers the impressive extant remains of buildings. Thus, it is extremely fortunate that the evidence of the mathematical papyri can be complemented by sources like papyrus Berlin 10005, the Reisner papyri, papyrus Harageh 3, and others.

---

42   See, for example, the depiction in the tomb of *Niankh-Khnum* and Khnum-hotep (Old Kingdom, Fifth Dynasty); also see Moussa/Altenmüller, *Nianchchnum*, pl. 23.

NEW KINGDOM

Many of today's most famous kings and queens of Egypt ruled during the New Kingdom (1550–1069 BCE), rendering it one of the best-known periods of ancient Egypt in modern times.[1] Because it is more recent than Egypt's previous two periods of bloom, it is not surprising that many more documents are extant. Monumental inscriptions enabled Egyptologists to get a fairly good picture of the political history in this time. Royal and private tombs have provided ample information about religion. Furthermore, due to the use of ostraca (stone or pottery shards) as cheap writing material, there is, besides the papyri, a second medium to record daily life issues. Ostraca are more robust than papyrus, and thus thousands of them are still extant and allow us better insights into daily life affairs. Although ostraca have been used from the Old Kingdom on, it is during the New Kingdom that they became more popular.

Within the New Kingdom, several subperiods can be distinguished. The New Kingdom begins with the expulsion of the Hyksos by Ahmose (1550–1525 BCE), the founder of the Eighteenth Dynasty. In the following reigns of this dynasty, the territory of Egypt was reestablished, and Egypt reaches its third cultural peak. This is the last period of pharaonic Egypt, in which, at least temporarily, Egypt held an influential political position. A major break during the New Kingdom was caused by Amenhotep IV, better known as Akhenaten, who strongly promoted the god Aton and moved the capital from Thebes to a place now known as Tell el Amarna. However, the fundamental changes that Akhenaten tried to bring about were short lived and already reversed by the immediately following rulers, Smenkhkara (about whom we know next to nothing), Tutankhamun (whom we only know because of his outstanding tomb), Ay, and Horemheb, the last kings of the Eighteenth Dynasty. The following Nineteenth and Twentieth Dynasties are also known as the Ramesside Period, a name that was derived

---

1   For a concise overview of history and politics of the New Kingdom, see Shaw, *History*, pp. 218–313.

from that of the majority of their kings, "Ramses." The Nineteenth Dynasty begins with Ramses I (1295–1294 BCE), and the Twentieth Dynasty ends with Ramses XI (1099–1069 BCE). Between these two kings, Egypt became, for the last time, a distinguished independent power, with its peak presumably during the long regency of Ramses II (1279–1213 BCE).

The wealth of evidence extant from the New Kingdom also includes a variety of information on the Egyptian officials during this time. In the later New Kingdom, the education of scribes was systematized, as evidence from the genre of school texts seems to indicate. A large variety of compositions were used as models, which were presumably copied by scribes for instruction.[2] Rather surprisingly, there are no extant mathematical handbooks from this period. The only two mathematical texts extant are two ostraca, both of which show no more than a few incomplete lines of text.[3] Despite this, there is still evidence within the school literature that mathematics remained a key feature of scribal education. In addition, ration lists and several other documents prove that mathematical techniques were still an important part within the daily life of a scribe. The areas of numerical responsibilities of the officials apparently remained unchanged.

The tomb of Menna, an Eighteenth-Dynasty official (probably during the reign of Thutmose IV), who held the office of a "scribe of the fields of the lord of the two lands," depicts the measuring of a size of a field (figure 24).[4] As in the earlier text of papyrus Harageh 3, this is executed not by a single person, but by a team, presumably for practical reasons as well as to prohibit any fraudulent measuring. The paintings of the tomb of Nebamun, the "scribe and grain accountant in the granary of divine offerings of Amun," include several scenes concerning the administration of goods—for example, agricultural scenes showing the harvest of grain and the viewing of the produce of the estates, in which animals of the estates are led before Nebamun by officials who also bring the corresponding documentation.[5]

The evidence of scribal life from this period comes mainly from one site, the town of Deir el Medina.[6] This town on the west bank of the Nile, near Thebes, used to house the workmen and artists who constructed and decorated the tombs in the nearby Valley of the Kings and the

---

2   Translations of school texts can be found in Lichtheim, *Literature II*, pp. 167–78, and Simpson, *Literature*, pp. 438–41.

3   Senmut 153 and Turin 57170.

4   The tomb is described in detail in Hartwig, "Tomb of Menna."

5   For a detailed description of the viewing of the produce of the estates, see Parkinson, *Nebamun*, pp. 92–109; for the agricultural scenes, see Parkinson, *Nebamun*, pp. 110–19.

6   A good introduction to various aspects of the evidence from this town can be found in Bierbrier, *Tomb-Builders*; Lesko, *Pharaoh's Workers*; McDowell, *Village Life*; and Meskell, *Private Life*. The ample evidence provided by Deir el Medina will certainly be exploited in further studies. For more information about ongoing research projects, see also the online database http://www.leidenuniv.nl/nino/dmd/dmd.html (accessed September 15, 2012).

FIGURE 24: Measuring the size of a field from the tomb of Menna (TT69)

Valley of the Queens. Like the Middle Kingdom town Lahun, Deir el Medina is not situated in the immediate proximity of the Nile, a fact that has led to its conservation and the possibility of its modern excavation. It is from this town that most of our knowledge about daily life during the New Kingdom comes, although it has to be noted that the community of Deir el Medina was by no means ordinary. Apart from numerous papyri, more than 10,000 ostraca have been found, and the majority of the sources discussed in the next chapter are likely to have come from this place.

12.

New Kingdom Mathematical Texts: Ostraca Senmut 153 and Turin 57170

Almost no mathematical texts from the New Kingdom are extant. So far, only two fragmentary ostraca have been published: Senmut 153 and Turin 57170. The Senmut Ostracon comprises an incomplete (possible) title, followed by six lines of computation:[1]

OSTRACON SENMUT 153

$[\dots]\ \overline{3}\ \overline{14}\ \overline{2}\ \overline{21}\ [\dots]$			
1	$\overline{7}$		
	3		
2	$\overline{6}$	$\overline{14}$	$\overline{21}$
	$3\,\overline{2}$	$1\,\overline{2}$	1
4	$\overline{2}$	$\overline{14}$	
	$10\,\overline{2}$	$1\,\overline{2}$	

The first line of the ostracon is too fragmentary to be reconstructed.[2]

Since no checkmarks nor a result of the calculation are preserved, it could be read as the multiplication of 4, 5, 6, or 7 times $\overline{7}$ or simply as the beginning of a multiplication table for 7.[3] The individual steps of the calculation are achieved with the aid of auxiliary numbers. The initial doubling may (also) have been aided by the use of the $2 \div n$ table and the auxiliary

---

1   Senmut 153 is published in Hayes, *Senmut*, pp. 29–30 and plate XXIX.
2   Hayes, *Senmut*, p. 29, reads . . . cubit, palm (?) 3 14 2 21 3 (?) and interprets this as "Problem: convert into cubits 3 palm."
3   See Fowler, *Mathematics of Plato's Academy*, p. 269.

number then introduced at a later stage. The auxiliary numbers use $\overline{21}$ as a reference (= 1). Consequently, $\overline{7}$ equals 3 (line 1 and 2 of the calculation) and $\overline{14}$ equals $1\overline{2}$.

OSTRACON TURIN 57170

[. . .]You shall say: 1 ḥ3r.

[. . .] You shall say: 2 ḥ3r.

[. . .] You shall say: 2 ḥ3r 1 ḥq3.t.

[. . .] You shall say: 2 ḥ3r 2 ḥq3.t.

[. . .]You shall say: [. . .].

The ostracon, today kept in Turin, likewise contains only a few (and incomplete) lines of text that seem to indicate solutions.[4] The attribution as a mathematical text is, nevertheless, secured by the form of the sḏm.ḥr.f.

---

4   For a facsimile of Turin 57170, see López, *Ostraca Ieratici*; for a translation, see Imhausen, *Algorithmen*, p. 363.

13.

Two Examples of
Administrative Texts

D espite the scarce (and rather disappointing) evidence of mathematical texts from the
New Kingdom, ample evidence from administrative texts ascertains the significance that
the knowledge of mathematical practices must have held in the work life of a scribe.[1] Numerous
accounts display the vital function that mathematics still had in running the administration of
rations, temple offerings, labor, produce, and others. According to Jeffrey Spencer, the accounts
of Seti I are "notable for the extraordinarily fine handwriting of even routine accounts compiled
in the royal chancellery," among which is "one of the finest examples of writing in the British
Museum collection."[2] Learned numeracy thus continued to be an essential part of scribal culture, and those proficient in it were proud of this ability. Consequently, scribes are still depicted
as accountants in tombs, as, for example, in the scene of measuring land from the tomb of
Menna (TT69, see figure 24) or the accounting scenes in the tomb of Nebamun.

As in earlier times, it is difficult to obtain information about actual mathematical practices from the accounts. Unlike the mathematical texts that state a mathematical situation,
explicitly indicate what the scribe should calculate, and then detail the procedure with which
it is done, the accounts indicate only numerical data, some of which are the result of mathematical operations that the scribe performed. However, even the accounts suffice to illustrate
basic mathematical concepts used during the New Kingdom, as the following two examples
demonstrate.

---

1   A selection of the most significant documents can be found with description and analysis in
     Warburton, *State and Economy*. On Egyptian administration in various contexts and periods see Garcia,
     Administration.
2   Spencer, *Ancient Egypt*, p. 183.

13.1 PAPYRUS HARRIS I

Papyrus Harris I (papyrus BM EA 9999), named after the man who purchased it (Anthony Charles Harris) in 1855, came to the British Museum in 1872.[3] At over 40 m length, it is one of the longest papyri extant from ancient Egypt. Papyrus Harris I is primarily a political document, documenting the relation between Ramses III and the temples of Egypt. It is divided into five sections of hieratic text and three illustrations of the king before the gods of Thebes, Heliopolis, and Memphis as the three religious centers.[4] The illustrations are accompanied by hieroglyphic texts. The hieratic parts consist of lists of temple endowments and also include a brief summary of the entire reign of King Ramses III. These endowments were not granted as gifts to the temples but were intended for the royal cult.[5] The amounts of endowments were enormous: The list of endowments relating to Theban temples alone includes 309,950 sacks of grain as well as large quantities of metals and semiprecious stones. In order to give an idea of this text, the introduction and a later part with some endowments are given in translation.[6]

Papyrus Harris I, Introduction

"(1,1) Year 32, third month of the Harvest season,[7] sixth day; under the majesty of the king of Upper and Lower Egypt; Usermaatra-Meryamun, alive, sound, and healthy; Son of Re: Ramses (III), Ruler of Heliopolis, alive, sound, and healthy, beloved of all gods and goddesses; (1,2) king, shining in the White Crown like Osiris; ruler, brightening the Netherworld like Atum; ruler of the throne of the great house in the midst of the necropolis, traversing eternity forever as king of the Netherworld; King of Upper and Lower Egypt; Usermaatra-Meryamun; Son of Re: Ramses (III), Ruler of Heliopolis, alive, sound, and healthy, the Great God (1,3) says in

---

3   After initial translations or partial translations and related publications by Eisenlohr, Birch, Piehl, and others (for an overview of the early publications, see Schaedel, *Listen*, pp. 11–13), the text was comprehensively edited by Pierre Grandet (*Papyrus Harris I*). For a hieroglyphic version of the text, see Erichsen, *Papyrus Harris I*. An English summary with selected excerpts in translation can be found in Warburton, *State and Economy*, pp. 194–216.

4   For a schematic overview of the overall structure of the text, see Grandet, *Papyrus Harris I*, p. 21, tableau 2.

5   Spencer, *Ancient Egypt*, p. 184.

6   The following translations are based on those of Breasted, *Records of Egypt*, and Grandet, *Papyrus Harris I*.

7   The Egyptian civil calendar used a division of the year into three seasons (*ȝḥt* Inundation, *prt* Growing, *šmw* Harvest) of 4 months each. The Egyptian year originally began around the time of the beginning of the inundation of the Nile (mid-July). Each month consisted of 3 weeks of 10 days each. At the end of the year, 5 extra days, the epagomenal days, were added. Thus, the Egyptian civil calendar had 365 days, that is, it was ¼ day shorter than the solar year. On the Egyptian calendar, see Parker, *Calendars*.

praise, adoration, and laudation, the many benefactions and mighty deeds, which he did as king as ruler on earth, in favor of the domain (*pr*) of his august father; Amun-Ra, king of gods, (1, 4) Mut, Khonsu, and all the gods of Thebes;

in favor of the domain of his august father, Atum, lord of the Two Lands; the Heliopolitan Re-Harakhte; Iousaas, mistress of Hetepet and all the gods of Heliopolis;

in favor of the domain of his august father, (1,5) Ptah, the great, South-of-His-Wall, lord of "Life-of-the-Two-Lands"; Sekhmet, the great, beloved of Ptah; Nefertem, defender of the Two Lands and all the gods of Memphis;

in favor of the august fathers, all the gods and goddesses of South (1,6) and North;

as well as the good benefactions [which he did for] the people of the land of Egypt and every land, to unite them all together; in order to inform (1,7) the fathers, all the gods and goddesses of South and North, and all [foreigners], all citizens, all (common) folk, and all people, of the numerous benefactions and many mighty deeds, (1,8) which he did upon earth as great ruler of Egypt."

This introduction sets the stage for the following endowments. Rameses III addresses the gods in three "divisions," those associated with Thebes (Amun-Ra, Mut, and Khonsu), those associated with Heliopolis (Atum, Re-Harakhte, Iousaas), and those associated with Memphis (Ptah, Sekhmet, and Nefertem), as well as all the gods of Upper and Lower Egypt. The last section of the introduction also includes the human sphere among the recipients of the benefactions.

From a mathematical point of view, the following endowments are the most interesting of this text, because they include lists of individual items that were part of these benefactions. Within these lists, the administration of items is apparent using various metrological systems, and there is also a summary of several items that were regarded as belonging to one group of things. As an example, an excerpt of one list from the section regarding Memphis, the third section within the Harris papyrus, is given shortly. The section begins with an illustration (figure 25) showing the king worshipping the gods of Memphis, namely, Ptah, Sekhmet, and Nefertem. The king holds the royal insignia crook and flail and is dressed with clothing exclusive to him. The accompanying hieroglyphic inscription includes the names of the persons involved written in a short column in front of their faces as well as two columns of text representing words that the king addresses to the gods. The illustration is followed first by an introduction containing a speech of the king addressed to the gods, and then by a description

FIGURE 25: Illustration from papyrus Harris I: Ramses III before the gods of Memphis

of the temple buildings and equipment that Ramses implemented at Memphis. Five lists detailing various items that were also dedicated then follow this section:[8]

"GOLD and silver, real lapis lazuli, real turquoise, every sublime precious stone, copper, black copper, garments of royal linen, of *mk*-linen, of fine linen of good quality, of fine linen, smooth garments, vases, bovine animals, birds and all things, which the king Usermaatre Meryamun—may he live, be prosperous and healthy—the great god gave as gifts of the lord, may he live, be prosperous and healthy, to the estate of Ptah, the great, the one who is south of his wall, lord of the life of the two lands, and to the temples which are attached to it from the year 1 to the year 31, making 31 years:

Perfect gold, gold of two times, white gold in the form of vessels and ornaments	*dbn* 263, 5 $\overline{2}$
Gold. An ornament of a magistrate	*dbn* 2
Silver as vessels and flat metal	*dbn* 342, 7 $\overline{6}$

8   See Grandet, *Papyrus Harris I*, pp. 294–96, and Breasted, *Records of Egypt*, vol. 4, pp. 172–74. My translation is based on the works of Breasted, Erichsen, and Grandet.

Silver in worked form:	1 large tablet of 1 cubit and 6 palms in width and of 1 cubit, 1 palm and 3 fingers in height, which makes in *dbn* 173, 8 $\overline{\overline{3}}\,\overline{6}$
TOTAL: silver as vessels and ornaments	*dbn* 516, 6
TOTAL: gold as vessels, ornaments and flat metal	*dbn* 780, 1$\overline{2}$
Real lapis lazuli, mounted in gold and attached with two strings to a necklace	1, which makes in *qdt* 3
Real lapis lazuli	*dbn* 3, 2
Real turquoise	*dbn* 2
Real green feldspar	*dbn* 10
Real lapis lazuli and turquoise	36 scarabs mounted and encased in gold
Lapis lazuli	46 large scarabs
Turquoise	46 large scarabs
Copper in worked form of an alloy of six parts	1 large tablet, which makes in *dbn* 245
Copper in worked form of an alloy of six parts	1 tablet, which makes in *dbn* 65
Copper as vessels and flat metal	*dbn* 1708
TOTAL: Copper as vessels and flat metal:	*dbn* 2018
Royal linen, *mk*-linen, fine linen of good quality, fine linen, colored linen	7026 garments
Myrrhe	*dbn* 1034
Fresh incense, honey, oil, fat, butter	1046 individual jars
šdḥ drink and wine	25978 individual jars
TOTAL:	27024 individual jars
Ivory	1 tusk
Storax-wood	*dbn* 725
Aleppo pine	*dbn* 894
Tj-špsy wood/produce	45 bundles
Qnnj and *Tj-špsy*	28 baskets
šrt-Grain from Syria	40 *ḥqꜣ.t*
Mint	40 baskets

Iwft-plant	80 baskets
Absinthe	11 baskets
Fruits	14 *ḥqȝ.t*
Pine	8 boards
Black eye paint	*dbn* 50
Rdm.t plant	50 mats
Natron	*dbn* 14400
Fayence	31000 pearls
Crushed fayence	441 *hnw*
Fayence	3200 *ḥtm* amulettes
Stuccoed wood	31 boxes (in the form of cartouches)
Oxen, calves, various cattle, oxen with short horns	
Total of bovine animals	*979*
Live geese	269
Live white-fronted geese	150
Live *wrdw*-geese with short beaks	1035
Live *wrdw*-geese	41980
Live waterfowl	576
TOTAL of various birds	44010"

The introductory section of the list just cited lists the kind of things that are then noted within the list with their respective quantities. The mathematical content of these lists is straightforward; within sections where individual items of one kind are listed, a total is indicated at the end of this section—for example, the various items of copper (copper in worked form of an alloy of six parts and copper as vessels and flat metal) are first listed with individual numbers and specifications (1 large tablet, which equals 245 *dbn*, 1 tablet, which equals 65 *dbn* and an amount of 1708 *dbn*), and then the total of the copper is indicated as 2018 (245 + 65 + 1708) *dbn*. Thus, the idea of using a specific criterion, such as the weight of an object, is used in order to make the object comparable to others and countable within an account of items of one kind.

Similarly, there are totals of various items (fresh incense, honey, oil, fat, butter, *šdḥ* drink and wine) delivered in jars and a total of a number of various birds. Within the lists, the quantities of the individual items are indicated in *dbn* (a weight unit; where it is followed by two numbers it indicates weight in *dbn* and *qdt*), in absolute numbers, in a capacity unit, or in other units that are appropriate for the individual things. There is no attempt to establish an overall number (e.g., the value) of the complete list.

As may be obvious from the previous sections, papyrus Harris I is an account of the achievements of Ramses III for various temples and, therefore, their gods, especially those detailed in the individual sections on Thebes, Heliopolis, and Memphis. The carefully executed illustrations and the elaborate introductions elevate the accounts that are given from the basic administrative level into the realm of a royal discourse with the gods. Ramses III uses the format of an account within this discourse, maybe to place emphasis on the material abundance that he bequeathed to the gods. At the same time, the use of an account within a text of this kind may also indicate the importance and value that were placed on Egyptian accounts in ordinary life. Establishing an account and determining the respective quantities or numerical values that are noted within this document constituted an act of ascertaining facts and proving certain achievements. Apparently, this was also a valid method in a communication between king and gods.

13.2 PAPYRUS WILBOUR

The Wilbour papyrus is an equally exceptional document in length and content as papyrus Harris I described in the previous section.[9] Although its absolute dimensions are somewhat smaller (papyrus Harris I has a length of 41 m) at about 10 m in length, the Wilbour papyrus is one of the few chance finds of reasonably well preserved larger papyri that hint at the role that administrative organisation held in pharaonic Egypt.[10] The topic of the Wilbour papyrus is the administration of arable land, and thus this text is one of the sources one should take into account when reading the famous statement of Herodotus that geometry was invented in Egypt and then passed into Greece.[11] The content of the Wilbour papyrus provides detailed insights into various aspects of agricultural administration in later New Kingdom Egypt, that is, the time of Ramses V (1147–1143 BCE).[12] However, as everyone who has focused on this

---

9   The edition of Gardiner, *Papyrus Wilbour*, includes photos (vol. 1), translation (vol. 2), commentary (vol. 3), and index (vol. 4). This edition has been followed by a small number of subsequent studies, namely, Janssen, "Agrarian administration" (this article is a commented summary of a monograph published in 1982 in Russian on the Wilbour papyrus by I. A. Stuchevsky, which would otherwise be unavailable for Western readers), Helck, *Materialien zur Wirtschaftsgeschichte*, Menu, *Régime juridique des terres*, and Katary, *Land Tenure in the Ramesside Period*. Because of its detailed information on charges that the cultivators had to pay, the Wilbour papyrus often features prominently in works on agrarian administration, economy, and taxation, such as Katary, "Taxation," Warburton, *State and Economy*, esp. pp. 165–69, and Janssen "Role of the temple," to name but a few.

10 Lengths according to Spencer, *Ancient Egypt*, p. 184 (Harris papyrus), and Gardiner, *Papyrus Wilbour*, vol. 2, p. 1 (Wilbour papyrus).

11 Herodot, *Histories*, Book II: 109. For a commentary see Lloyd, *Herodotus Book II*.

12 The few other, earlier, but also much smaller sources for the administration of arable land are the Eighteenth Dynasty Asmolean 1984.61.I papyrus (published in von Lieven, "Fragmente eines Feldregisters") and the fragmentary leather roll of the Nineteenth Dynasty, which details an "assessment of plots of land held by various smallholders at the uniform rate of 2½ sacks of corn per aroura" (Gardiner, Ramesside administrative documents, p. xx). Three related sources from the Twenty-first

text has noted, much of the intricacies of Egyptian land administration that are obviously implicated in this papyrus remain unclear to us.[13] The following paragraph attempts to summarize the results of those who have intensely studied this document (Gardiner, Stuchevsky, Menu, Janssen, Katary, and Warburton).

The Wilbour papyrus consists of two separate texts on its recto and verso, which are usually referred to as Text A and Text B in secondary literature. Text A lists more than 2800 plots of land, which were organized according to their administrating institutions, mostly temples but also secular institutions.[14] Geographically, these plots were located in Middle Egypt, from Medinet el Fayum to Minia.[15]

The text indicates their respective locations and dimensions and gives an assessment of the individual plots, that is, the amount of grain, which the cultivator (who is also listed) was supposed to deliver.[16] Cultivators include temple agents as well as small farmers working for themselves, but also priests and soldiers to high-ranking officials like the vizier (the latter employing laborers).[17]

As an example, paragraphs 7–9 of Text A are given here (the abbreviation mc. was introduced by Gardiner and indicates "measure of capacity"):[18]

§7   [The Sunshade of Re-]Harakhti which is in Ninsu.

      [Measure]ment made in the Pond of Iia's Tomb:

      [Land cultiva]ted by the cultivator Ramose *10*, mc. *5*, mc. *50.*

      [Measure]ment made in the Awen-grove south of Ninsu:

      [Land culti]vated by him *5*, mc. *5*, mc.*25.*

    Dynasty are Reinhardt, Parchov papyrus, and the Grundbuch papyri (see Gasse, *Données nouvelles*, and Vleeming, *Papyrus Reinhardt*).

13   As Janssen, "Agrarian administration," p. 352, put it: "Although these studies have enlarged our understanding of various aspects of the text, many problems as yet remain unsolved. Even the main purport of the papyrus is still under discussion. That in the almost forty years since the publication of Gardiner's Commentary so little has been achieved in this respect proves, on the one hand, the outstanding quality of this scholar's work; on the other, it attests to the intricate problems which this administrative document poses to us."

14   Katary, "Land-tenure," p. 62.

15   For a map indicating the various locations of these plots, see Katary, "Land-tenure," p. 63 and p. 64.

16   See Katary, "Taxation," pp. 9–10. On the term cultivator (*jḥwtjw*), see Janssen, "Agrarian administration," p. 354.

17   Kemp, *Ancient Egypt*, p. 191.

18   Translation by Gardiner, *Papyrus Wilbour*, vol. 3, pp. 8–9. The use of red ink is indicated by the use of SMALL CAPITALS if it is text or italics in the case of numbers. Italic text indicates a word of Egyptian origin, such as *khar* (*ḫꜣr*). The capacity units were indicated as a general mc (measure of capacity) in Gardiner's translation. For their specific meaning, see Gardiner, *Papyrus Wilbour*, vol. 2, pp. 61–65, Warburton, *State and Economy*, p. 475, and Pommerening, *Hohlmaße*, p. 146. Gardiner's 2/4, which represent the Egyptian writing in submultiples, are rendered in the preceding translation in their respective value (1/2).

§8 [THE] MANSION of Ramesse-miamun who hears prayer in the House of Arsaphes.
MEASUREMENT made in the Lake of Perre:
Land cultivated by him *10*, mc. *5*, mc. *50*.

§9 THE MANSION of Ramesse-miamun in the House of Arsaphes.
MEASUREMENT made (to) the south-east of Tent-ioor:
Land cultivated by the cultivator Ramose 5 *5*, mc. 5 *5*, mc. 25 *25*.
MEASUREMENT made in the New land of T-gemy:
Land cultivated by him *10*, mc. *10*, mc. *100*.
Another *10*, mc. *7 ½*, mc. *[7]5*.
Another *20*, mc. *5*, mc. *100*.
MEASUREMENT made in the island west of . . . nukhesh:
Land cultivated by him *10*, mc. *7 ½*, mc. *7[5]*.

This example suffices to give a most general idea of the first part of the Wilbour papyrus, the wealth of information that it contains but also the difficulties that we are as yet unable to solve. Each paragraph begins with the statement of the name of the respective institution—many of which are not known to us from other sources. This is followed by indications of the location of the plots of land that were measured. These indications use landmarks and the indication of cardinal points and can have given only a rough indication of where the respective plot of land was located. Hence, these indications were not meant to provide sufficient information to locate the plot for someone who did not already know where it was. Rather, it indicates which plot was meant to those who knew where the property of the respective institution was located. Then the sizes of the plots follow, which are linked to an amount of grain (per unit) and a total amount of grain (resulting from the multiplication of size and amount of grain per unit). Also mentioned are those who are responsible for it.

Thus, in section 7, the first line indicates the institution by its designation, "The Sunshade of Re-Harakhti" (i.e., a solar shrine), with its location, Ninsu (Herakleopolis Magna). Two plots are then listed by their respective geographical landmarks, that is, the Pond of Iia's Tomb and the Awen-grove south of Ninsu. Both plots were associated with a certain Ramose. The numeric values associated with these plots were their sizes in arouras (i.e., 10 and 5) as well as the multiplicative number that links the plots with a quantity of grain per aroura (i.e., 5 in both examples); finally, the respective amount of grain for the fields is indicated, which can be calculated as the product of size and the indicated multiplicative number (e.g., in the first of the plots, the size of the plot is indicated as 10 arouras and the number indicating the amount of grain per aroura as 5 mc., resulting in an amount of 50 mc. for the total field; in the second example, the size of the plot is given as 5 arouras, the number indicating the amount of grain per aroura as 5 mc., resulting in an amount of 25 mc. for the second field).[19]

The scribe of the Wilbour papyrus used red ink elaborately to structure the text, as can be seen from the three preceding paragraphs. Red is used to mark the beginning of a paragraph, that is, a new institution. In all the preceding examples, only the beginning of the designation of the institution is marked in red (e.g., in section 9, the only paragraph, where the complete line is extant: "THE MANSION of Ramesse-miamun in the House of Arsaphes"). On a second level, the individual plots that belong to one institution are also marked by an initial rubrum. The general structure indicates the geographic specification of the plot first, followed by the name of the cultivator and then the three numeric values, size of the field in arouras, quantity of grain per aroura, and quantity of grain for the respective field. These three numbers are also usually written in red, although in some cases they were (presumably erroneously) written in black (e.g., in the case of the second plot of section 7), in some cases the black numbers were supplemented by (identical) red numbers (e.g., in the case of the first plot of section 9). Thus, the usage of the red ink highlights the essential information of this administrative document, so it can easily be grasped.

---

19   See Gardiner, *Papyrus Wilbour*, vol. 2, p. 62.

14.

Mathematics in Literature

In addition to the accounts, evidence for the significance of mathematics for scribal culture also comes from some literary texts, most notably, the *Late Egyptian Miscellanies*.[1] The *Miscellanies* are a group of various texts, including model letters, excerpts of literary compositions, praises of the scribal profession, eulogies, and hymns to a teacher and others. During the Ramesside Period, scribes copied them on a regular basis.[2] Several of these texts refer explicitly to the profession of the scribe, its demands and rewards. The so-called *Satire of the Trades*, extant in several versions, compares the duties and privileges of a scribe with those of other professions (e.g. that of a farmer, a soldier and others). Direct or indirect references to mathematics can be found frequently within these sources as the following examples illustrate.

14.1 MATHEMATICAL EDUCATION

Papyrus Anastasi V, 22,6–23,7

"I have placed you at the school along with the children (22,7) of magistrates in order to instruct and to teach you concerning this aggrandizing calling. Look, I tell (22,8) you the way of the scribe in his (perpetual) 'Early to your place! Write in

---

1   For an overview and discussion of these texts see Hagen, "Late Egyptian Miscellanies." For an English translation, see Caminos, *Late Egyptian Miscellanies*; a more recent German translation can be found in Tacke, *Verspunkte*; Simpson, *Literature*, pp. 438–41 provides translations of some selected examples.

2   Originally, the miscellanies were interpreted to have been the result of the schooling of junior scribes; however, this assumption of Adolf Erman (see Erman, *Altägyptische Schülerhandschriften*) was made at a time when not all the manuscripts were properly published (Hagen, "Late Egyptian miscellanies," p. 84). At present, it seems more likely to assume them to have been part of a body of texts that were perceived by the scribes as inherent to their culture.

front of your companions! Lay your hand (23,1) on your clothes and attend to your sandals!' You bring your book daily with a purpose, be not (23,2) idle. They (say): 'Three plus three.' You ... another happy occasion in which you grasp the meaning of a papyrus-book. (23,3) Make 15 mats more ... lifetime, and begin to read a dispatch. You will (23,4) make calculations quietly: let no sound [coming forth (?)] from your mouth be heard. Write with your hand, read with your mouth, and take advice. (23,5) Be not weary, spend no day of idleness, or woe to your limbs! Fall (23,6) in with the ways of your instructor and hear his teachings. Be a scribe. 'Here am I', thus shall you say as often as (23,7) [(I) call] to you. Beware of saying ... !"[3]

This section is an example of a letter from a father to his son. The father wants to ensure that the son takes his studying seriously. The admonitions begin by telling the son to be punctual and to appear with orderly clothing. The tasks that the student is meant to perform include an explicit reference to mathematical education ("Three plus three."). The addition mentioned in this example may be a placeholder for all kinds of mathematical operations that the scribes were trained to perform. During the lessons, textbooks in form of papyrus rolls were used ("you grasp the meaning of a papyrus-book")—and it is exactly in this context in which we imagine a text like the Rhind or Moscow papyrus was used. Furthermore, we learn about discipline being taught to junior scribes; for example, they were supposed to obey their teachers ("Fall in with the ways of your instructor and hear his teachings") and, with regard to their conduct in their mathematics education, to perform their calculations in silence ("You will make calculations quietly: let no sound from your mouth be heard"). Punishment for misconduct was physical ("woe to your limbs").

Papyrus Anastasi V, 15,6–17,3 = Papyrus Sallier I, 6,1ff

(15,6) To the following effect: I am told that you have abandoned writing and whirl around in pleasures, that you have applied yourself to working (15,7) in the field and have turned your back upon the god's words. Have you not recalled the condition of the cultivator faced with the registration (16,1) of the harvest-tax after the snake has carried off one half and the hippopotamus has eaten up the rest? The mice abound (16,2) in the field, the locust descends, the cattle devour. The sparrows bring want (16,3) upon the cultivator. The remainder that is on the thrashing platform is

---

3   Translation from Caminos *Late Egyptian Miscellanies*, pp. 262–63.

(almost) at an end, and is for the thieves, whilst (16,4) the value of the hired cattle is lost. The yoke (of oxen) being dead on account of (too much) threshing and
ploughing. (16,5) (Now) a scribe has landed at the river bank and is about to register
the harvest tax; the apparitors (16,6) carrying staffs and the Nehsyu rods of palm.
They say: 'Give corn!', though there is none. (16,7) They beat (him) furiously. He
is bound and thrown into the well; he is soused in a headlong dipping. (16,8) His
wife has been bound in his presence, his children are in fetters. (17,1) His neighbors
abandon them and are fled. (Thus) their corn flies away. But (17,2) a scribe, he is
a controller of everyone. He who works in writing is not taxed, he has no dues (to
pay). (17,3) Take note of this."[4]

This is another section from a letter (presumably of a parent) to a student scribe, in which
the student is admonished to study diligently. In contrast to the previous letter (Anastasi V
papyrus, 22,6–23,7), the situation depicted here is that of a student who has (temporarily)
left his studies, apparently to enjoy himself while working in a field. In order to strengthen
the admonition, the student is reminded of the fate of a peasant (i.e., the occupation that he
is currently enjoying) at harvest time. This is then compared to the working life of a scribe.
The scribal profession is depicted as far more pleasant, first of all within the description of
the peasant's suffering when the scribe is the one to inflict it. The explicit reference to the
scribe at the end of the letter outlines his superior position: he has no dues to pay and hence
can never get into the unfortunate situation of the peasant. Note that the function the scribal
profession has in this letter is that of accounting—that is, exactly for what the scribe needs his
mathematical education.

Papyrus Lansing, 8,7–10,10

To the following effect: (8,8) Behold, I am teaching you and making sound your
body to enable you to hold the palette freely, to cause you to become (8,9) a trusty
one of the king, to cause you to open treasuries and granaries, to cause you to receive
(corn) from the ship at the entrance of the granary, (8,10) and to cause you to issue
the divine offerings on festal days, attired in (fine) raiment, with horses, whilst your
bark is on (9,1) the Nile, and you are provided with apparitors, moving freely and
inspecting. A villa has been built in your city, and you hold (9,2) a powerful office,

---

4   Ibid., p. 247.

by the king's gift to you. Male and female slaves are in your neighborhood, and those who are in the fields in holdings of your own making will grasp your hand. Look, (9,3) I am putting you to be a dependent of life. Put writing in your heart that you may protect yourself from all manner of toil and be a worthy magistrate. [. . .] "[5]

Once more a scribe is reminded in a letter to work diligently. But this time, the reward for his hard work is set in front of him. As a scribe he will be prosperous and powerful—and enjoy a high standard of living. Again, all the functions described require mathematical knowledge: the scribe is responsible for the administration of treasuries and granaries, the receipt of taxes (in form of grain deliveries), and the administration and distribution of divine offerings.

14.2 MATHEMATICS IN THE SCRIBE'S DAILY WORK

The comparison between the profession of a scribe and other professions is a theme, which is often encountered in this type of text. And, naturally, the scribal profession is always depicted as being superior to the others. The reason given for this superiority is, as in the previous example, usually explicitly stated—the fact that scribes are literate and numerate and thus play a decisive role in accounting the produce of the other professions.

Papyrus Lansing, 4,2–5,7

Look for yourself with your own eye. The professions are set before you. The washerman spends the whole day going up (4,3) and down, all his [body] is weak through whitening the clothes of his neighbors every day and washing their (4,4) linen. The potter is smeared with earth like a person one of whose folk has died. His hands and (4,5) his feet are full of clay, he is like one who is in the mire. The sandal maker mixes tan; his odor (4,6) is conspicuous; his hands are red with madder like one who is smeared with his (own) blood and looks behind him for (4,7) the kite, even as a wounded man whose live flesh is exposed. The . . . prepares floral offerings and brightens ring-stands; (4,8) he spends a night of toil even as one upon whose body the sun shines. The merchants fare downstream and upstream, (4,9) and are as busy as brass, carrying goods from one town to another and supplying him that has not, although the tax-people (4,10) carry gold, the most precious of all minerals.

---

5   Ibid., pp. 400–2.

The ships' crews of every (commercial) house have received their load(s) (5,1) so that they may depart from Egypt to Djahy. Each man's god is with him. Not one of them (dares) say: 'We shall see Egypt (5,2) again'. A carpenter, the one who is in the shipyard, carries the timber and stacks it. If he renders today his produce of (5,3) yesterday, woe to his limbs! The shipwright stands behind him to say to him (5,4) evil things. His outworker who is in the fields, that is tougher than any profession. He spends the whole day (5,5) laden with his tools, tied down to his tool-box. He goes back to his house in the evening laden (5,6) with the tool box and the timber, his drinking mug and his whetstones. (5,7) But the scribe, it is he that reckons the produce of all those. Take note of this.[6]

The comparison focuses on the amount of physical labor and dirt, which is involved in the individual professions: the body of the washerman is worn out, the potter is covered in dirt, the sandal maker smells, and so on. As is explicitly stated in the case of a carpenter, a certain amount of work had to be delivered within a certain time to avoid painful repercussions. The scribe is depicted in a controlling function of all other professions as the one who numerically administers their work ("reckons the produce of all these"). It is the mathematical part of his work that puts the scribe in this superior position.

Papyrus Leyden 348 (vs. 6,1–vs. 9,6)

(vs. 6,1) The scribe Kawiser greets his lord the scribe Bekenptah. In life, prosperity and health! This letter is for my lord's information. (vs. 6,2) Another message to my lord: the house of my lord is well, his cattle (vs. 6,3) which are in the estate of my lord are well, his servants are well, and his cattle, which are in the field (vs. 6,4) are well. Do not be anxious about them. In life, prosperity and health! This letter is for my lord's information. Another message to (vs. 6,5) my lord: I have received the letter which my lord sent to (me) to say: 'Give corn-rations to the (vs. 6,6) soldiers and the Apiru who are dragging stone to the great pylon of . . . (vs. 6,7) Ra-messe-miamun (l.p.h.) 'Beloved of Maat' which is under the authority of the chief of Medjay Amenemone.' (vs. 6,8) I am giving them their corn-rations every month according to the manner (vs. 7,1) which my lord told me. [. . .][7]

---

6   Ibid., pp. 384–85.
7   Ibid., pp. 491–93.

In this example, the scribe copied a model letter, which details some of the possible duties of a scribe. They include, obviously, the distribution of rations, examples of which were found in the Middle Kingdom mathematical texts. "The manner which my lord told me" may refer to exact quantities or specific ratios for the individual workers and officials according to their rank, as seen in the earlier problem texts.

After all these eulogies of the scribal profession, it may be the time for a brief "reality check." Although, theoretically, the profession of a scribe was superior to all others, caution should be exercised in trusting these literary sources too much. Without doubt, the majority of scribes were employed in a modest social position (superior to farmers, but still not quite among the elite of the hierarchy), serving as accountants of temples or the state. The depiction of a typical scene of accounting can be seen in the decoration of the tomb of Amenhotep-Sise, a second prophet of Amun from the time of Thutmosis IV.[8] It is from these depictions that we get to see the "dark side" of the scribal profession, that is, what happened if a scribe failed to deliver his work. Despite the literary statement that the scribe is the one who administers the *b3kw* of other professions, he himself has *b3kw* to deliver as well, presumably in the form of "output in writing."[9] And his accounting, of course, needs to be found in order by his superiors. There are vivid graphic illustrations in tombs depicting what happens if the accounts do not add up, resulting in physical punishment.[10] Likewise, there is always the possibility that a scribe may lose his position and hence his livelihood, as indicated at the end of the literary letter of papyrus Anastasi I.

14.3 MATHEMATICS IN PAPYRUS ANASTASI I

Papyrus Anastasi I is a satirical letter; its modern editors date it to the Nineteenth Dynasty.[11] The text is extant in several sources of varying completeness. The main source is papyrus Anastasi I, after which the composition is named today.[12]

---

8   TT75 in Sheikh Abd el-Qurna. The tomb is published in Davies, *Tombs of Two Officials.*
9   Warburton, *State and Economy*, interprets *Papyrus Anastasi V*, 17, 2 as an explicit attestation of this duty. For a discussion of the term *b3kw* and its attestations, see Eyre, "Work and organization of work," p. 193, Römer, *Gottes- und Priesterherrschaft*, pp. 382–85, and Warburton, *State and Economy*, pp. 237–60.
10 The accounting scene from the tomb of Amenhotep-Sise is depicted in Loffet, *Scribes*, Doc. XVIIIe Dynastie 15, p. 347 (scene before p. 347). The bottom register shows a person lying on the ground and several persons with sticks in hand standing around him.
11 There have been two editions so far. The first edition, Gardiner, *Egyptian Hieratic Texts*, was followed by a compilation of available sources (Fischer-Elfert, *Anastasi I-Text*) and then a German edition, Fischer-Elfert, *Anastasi I-Übersetzung*, taking into account the new sources available since 1964. For a discussion of the date of Anastasi I papyrus, see Fischer-Elfert, *Anastasi I-Übersetzung*, pp. 261–67.
12 So far, there are four more papyri and more than 70 ostraca known to supplement *Papyrus Anastasi I*. For a list of these, see Fischer-Elfert, *Anastasi I-Text*, pp. 1–4.

The letter is set to come from the context of a competition between two scribes, Hori, the sender of this letter, and Amenemope (Mapu), the recipient of this letter and the sender of a previous letter (of which no copies have been found yet, if it existed) to which this one is the reply. The text begins with an introduction of Hori, praising his abilities and detailing his competence as a scribe as well as his being part of the military. Hori then indicates the recipient of the letter, complete with praising his abilities. This is followed, as typical in Egyptian letters, by some wishes for the worldly existence as well as the existence in the afterlife of the recipient. Having completed this letter formula, the actual letter begins with the report of having received a letter from his colleague Amenemope (the recipient of this letter), which was (according to Hori) rather confused and not at all following the formal standards of letters. Hori concludes that Amenemope is indeed incapable of composing a letter at all and thus has summoned seven scribes and ordered them to write individual sections of this letter, which were then put together. This letter was supposed to bring out Hori's incompetence, challenging him with several scribal tasks, such as mathematical problems. However, the letter is such a mess that it is impossible to make much sense of it. Hori summarizes some of the tasks put to him (as far as he could make sense of the confused writing that he received), including a mathematical problem concerned with the extension of a lake, the administration of rations for soldiers, and a mathematical problem of determining the exact deficiency of a corn measure that is too small. The problems of digging a lake and the administration of corn rations will be resumed in the second part of Hori's letter and will be analyzed in there. The references to both these tasks are rather general. This is followed by the quote of a related problem (apparently also found in Amenemope's letter), to determine by how much a corn measure is too small (papyrus Anastasi I, 6,7–6,8):[13] How much is missing in one *ḥq3.t*, if the loss is 5 *hnw* for every *jp.t*?[14] The problem is stated like an example from the mathematical texts with actual data. Although Hori himself does not bother to indicate a solution, it is possible to give one in the style of the Middle Kingdom mathematical papyri:[15]

---

13   A reference to fraud involving false measuring devices can be found in the *Instruction of Amenemope*, chapter 16: "Do not tilt the scale nor falsify the weights, nor diminish the fractions of the grain measure" (Simpson, *Literature*, p. 236). A real case about fraud of this kind can be found in the Ostracon Leipzig 2 (Černý and Gardiner, *Hieratic Ostraca*). The workers complain that the measure used for determining their rations showed a shortage of several *hnw*.

14   For the varying sources of this passage, see Fischer-Elfert, *Anastasi I-Text*, p. 66. In *Papyrus Anastasi I*, the actual number indicating the shortage of the measure is left out. However, two other sources (Ostracon Deir el Medina 1177 and Černý and Gardiner, *Hieratic Ostraca* CXI) both show the 5 before *hnw*.

15   A modern solution involving algebraic equations can be found in Fischer-Elfert, *Anastasi I-Übersetzung*, p. 63.

Multiply 1 *jp.t* times 4, as 1 *jp.t* equals 4 *ḥq3.t*. 4 shall result.

Divide 5 by 4. 1 $\overline{4}$ shall result.

See, what is missing is 1 $\overline{4}$ *ḥq3.t*.

Instead of detailing Amenemope's tasks (or, behold, trying to solve them), Hori replies with a letter of his own, a letter like the one Amenemope meant to write had he only been capable of it. In this letter, he also includes various tasks a good scribe should be able to fulfill. The section of these tasks begins with a series of mathematical problems, which are analyzed in greater detail in the next section. This is followed by tasks concerning geographical knowledge. All these are accompanied by a prediction of Hori that Amenemope will not be able to cope with any of them and the dire consequences this will have for him.

As will be seen from the following extracts, the data provided are generally not sufficient to actually solve the indicated mathematical problems. Thus, rather than being taken from an actual mathematical textbook, the problems quoted in this letter were more likely meant to recall the style of mathematical education.

14.3.1 Digging a Lake (Papyrus Anastasi I, 13,6–13,7)

Hori begins his section on mathematical problems by referring to Amenemope's boasting and announcing that he will demonstrate his incompetence:

> Another topic: Look, you come here and fill me with your office. I will let you know your condition when you say: I am the commanding scribe of the soldiers. It has been given to you to dig a lake. You come to me to ask about the rations of the soldiers. You say to me: Calculate it! I am thrown into your office. Teaching you to do it has fallen upon my shoulders.

The digging of a lake is already mentioned earlier, when Hori summarizes the mathematical problems in Amenemope's letter (papyrus Anastasi I, 6,3–6,4):

> The sixth (scribe) runs off to measure the lake. He shall quadruple it in cubits, so that it can be dug out.

Apparently, the lake was supposed to be extended to four times its size. In order to determine this new size (and consequently the men needed to carry out this work, presumably in a given time), the actual size of the lake has to be determined first. This is the task that the sixth scribe sets off to fulfill.[16] Hori introduces a new aspect to this problem, that is, that of calculating the rations of the workers involved in the project. It will be seen that most of Hori's problems are related to the respective work forces that are needed to complete the tasks mentioned.

14.3.2 Constructing a Brick Ramp (Papyrus Anastasi I, 13,8–14,8)

The next section is a problem about a brick ramp.[17] Hori begins by pointing out Amenemope's position as a scribe—indicating that the following mathematical problem falls perfectly well in his duties:

> I will cause you to be embarrassed, I will explain to you the command of your master —may he live, prosper and be healthy. Since you are his royal scribe, you are sent under the royal balcony for all kinds of great monuments of Horus, the lord of the two lands. Look, you are the clever scribe who is at the head of the soldiers. A ramp shall be made of (length) 730 cubits, width 55 cubits, with 120 compartments, filled with reeds and beams. For height: 60 cubits at its top to 30 cubits in its middle, and an inclination of 15 cubits, its base 5 cubits. Its need of bricks shall be asked from the overseer of the troops. All the assembled scribes are lacking one who knows them (i.e., the number of bricks). They trust in you saying: You are a clever scribe my friend. Decide for us quickly. Look your name has come forward. One shall find one in this place to magnify the other thirty. Let it not be said of you: there is a thing that you don't know. Answer for us its need of bricks. Look, its measurements are before you. Each one of its compartments is of 30 cubits (in length) and a width of 7 cubits.

In order to perform the actual calculations needed to determine the number of bricks, the modern reader faces a number of difficulties that complicate the understanding of the given mathematical task.[18]

---

16   An example of an actual construction of a lake is reported in an inscription on several scarabs from the time of Amenophis III (1390–1352 BCE): "His majesty commanded the making of a lake for the great Queen *Tiye* in her town-quarter of *Djarukh*—its length being 3700 cubits, its breadth being 600 cubits" (Davies, *Historical Records*, p. 36).

17   No mathematical problems involving bricks are attested from any of the Middle Kingdom mathematical papyri. For the mathematical handling of brickwork in Mesopotamia in the Old Babylonian period (2017–1595 BCE), see Robson, *Mesopotamian Mathematics*, pp. 57–92.

18   For a detailed discussion of the terminology involved, see Fischer-Elfert, *Anastasi I-Übersetzung*, pp. 124–32. My summary relies on the assumptions detailed by Fischer-Elfert.

The ramp is described by its length (730 cubits), width (155 cubits), and height (60 cubits). In addition, it is mentioned that the height at the middle is 30 cubits—presumably to indicate that the ramp had a constant inclination. These data are sufficient to specify the basic wedge-shaped structure. Furthermore, 120 compartments are mentioned. This details the internal structure of the ramp. It was supposed to be constructed as a brick skeleton around these compartments filled with reeds and beams. The compartments are further specified as being of a length of 30 cubits and a width of 7 cubits. Also, a slope (or sloping side?) of 15 cubits and a base of 5 cubits are mentioned. This leaves the modern reader with the task of establishing the actual structure of this ramp and then with designating the indicated values to its individual parts. According to Reineke and Fischer-Elfert, the measurement given for the slope is given as a sum of both sides.[19] Thus, the actual slope measures only 7.5 cubits (on each side). This can be calculated into the respective width of the sloping side, which would increase with increasing height o the ramp. Alternatively (note that the former *terminus technicus sqd* is not used in the text of papyrus Anastasi I), the 7.5 cubits may designate a constant width of the sloping side of the ramp, which would then have a varying inclination. Theoretically, it is also possible that the 7.5 cubits indicate the *sqd* of the ramp at its highest point. However, this would result in a width of the sloping side at this point of 450 cubits, thus excluding this possibility.

The size mentioned for the compartments and their number do not agree with the total length of the wall. Therefore, Reineke assumed that the number was supposed to be 60 rather than 120. Fischer-Elfert has provided a calculation of the amount of bricks according to this structure and known brick sizes of the New Kingdom. He established that a total of 14,038,578 bricks were needed for the ramp.[20]

14.3.3 Transport of an Obelisk (Papyrus Anastasi I, 14,8–16,5)

Obelisks (Egyptian *ṯḥn*) are attested in Egypt from the Fifth Dynasty on.[21] They are made from a single stone in the shape of a steeply inclined pillar with a square base, and a small pyramid (pyramidion, Egyptian *bnbn(t)*) on top. The most common material used for their

---

19 Fischer-Elfert, *Anastasi I-Übersetzung*, p. 125 and Reineke, "Ziegelrampe," p. 7.
20 Fischer-Elfert, *Anastasi I-Übersetzung*, p. 132.
21 Representations of obelisks in various sources throughout Egyptian history are collected and discussed in Martin, *Garantsymbol*. For the technical aspects of obelisks, see Arnold, *Building*, pp. 36–40 and 67–70, and Rossi, *Architecture and Mathematics*, passim. For an overview of extant Egyptian obelisks, see Habachi, *Obelisks*, and now also the revised (German) Habachi and Vogel, *Obelisken*. For a detailed account of a modern erection of an obelisk, see Dibner, *Moving the Obelisks*. The unfinished obelisk left in the quarry at Aswan has been essential in the study of Egyptian techniques of manufacturing for obelisks;

manufacture was granite. Obelisks were connected with the solar cult and thus often found with temples. During the New Kingdom, they were often erected as pairs in front of a temple. The tallest extant obelisk is the Lateran obelisk of Thutmosis III, measuring more than 32 m. An unfinished obelisk left at the quarry in Aswan measured more than 40 m. The manufacture and transport of an obelisk are described in the decorations of the temple of Hatshepsut at Deir el Bahri.[22] The transport of an obelisk is also featured in the following section of Hori's letter:

> Hey Mapu, vigilant scribe, who is at the head of the soldiers, distinguished when you stand at the great gates, bowing beautifully under the balcony. A dispatch has come from the crown-prince to the area of Ka to please the heart of the Horus of Gold, to calm the raging lion. An obelisk has been newly made, graven with the name of his majesty may he live, be prosperous, and healthy of 110 cubits in the length of its shaft, its pedestal of 10 cubits, the circumference of its base makes 7 cubits on all its sides, its narrowing towards the summit 1 cubit, its pyramidion 1 cubit in height, its point 2 digits. Add them up in order to make it from parts. You shall give every man to its transport, those who shall be sent to the Red Mountain. Look, they are waiting for them. Prepare the way for the crown-prince Mes-Iten. Approach; decide for us the amount of men who will be in front of it. Do not let them repeat writing while the monument is in the quarry. Answer quickly, do not hesitate! Look, it is you who is looking for them for yourself. Get going! Look, if you hurry, I will cause your heart to rejoice. I used to [. . .] under the top like you. Let us fight together. My heart is apt, my fingers listen, they are clever, when you go astray. Go, don't cry, your helper is behind you. I let you say: There is a royal scribe with Horus, the mighty bull. And you shall order men to make chests into which to put letters that I will have written you secretly. Look, it is you who shall take them for yourself. You have caused my hands and fingers to be trained like a bull at a feast until every feast in eternity.

This section begins by pointing out the social position of the addressee, directly followed by the next task. Obviously, a scribe of the position indicated should be able to solve this

---

see Engelbach, *Problem of the Obelisks*. For a historical overview, see also Curran, Grafton, Long, and Weiss, *Obelisk*.

22　For a depiction of the obelisk on a ship, see Naville, *Deir el Bahri VI*, pl. CLIV. For a possible interpretation of these depictions and other aspects of handling obelisks, see Wirsching, *Obelisken Transportieren*.

without any problem. An obelisk has been made at Gebel Ahmar and is supposed to be transported from there, presumably to the residence.[23] Amenemope should determine the number of workers that are necessary for this transport. As Fischer-Elfert has pointed out, Hori's letter also contains a rough instruction of how to solve this problem:[24] the volume of the monument should be determined first, and from this result, the number of men needed for its transport can be calculated. Therefore, the dimensions of the obelisk are indicated in detail. The actual calculation of the volume, however, is not as straightforward as it might seem—again, the measurements given present us with some problems. The volume of the base can easily be calculated as the product of its three dimensions (1000 cubic cubits). The actual obelisk is said to have a lower base of 7 cubits, the upper side is said to be 1 cubit less on each side (5 cubits), and the shaft length is given as 110 cubits.[25] Thus, the volume can be calculated as in problem 14 of the Moscow papyrus, resulting in $3996\,\overline{\overline{3}}$ cubic cubits.

The top of the obelisk seems to be composed of two parts, the pyramidion of height 1 (and apparently base length 6 cubits) and another part called *ḥwy* or *ḥtr* of 2 digits, presumably indicating something on top of the pyramidion. As Fischer-Elfert noticed, the pyramidion, according to these measurements, is too flat in comparison with extant pyramidia.[26] He therefore assumed that the height was intended to be more than 1 cubit. Fischer-Elfert interprets the *ḥwy/ḥtr* as a coating that was supposed to cover the whole of the pyramidion.

14.3.4 Erection of a Colossal Monument (Papyrus Anastasi I, 16,5–17,2)

The following problem is again about a building project. Apparently, a monument is put up with the aid of sand; the calculations that the scribe is asked to perform are about the workmen it will take to complete the task in a given amount of time:

> You are told: Empty the magazine that has been loaded with sand under the monument for your lord—may he live, be prosperous and healthy—which has been brought from the Red Mountain. It makes 30 cubits stretched upon the ground with a width of 20 cubits, passing chambers filled with sand from the riverbank. The walls of its chambers have a breadth of 4 to 4 to 4 cubits. It has a height of 50

---

23 Gebel Ahmar (Red Mountain) was located near Heliopolis on the banks of the Nile.
24 Fischer-Elfert, *Anastasi I-Übersetzung*, p. 137.
25 Note that this length is more than double of the largest extant obelisk!
26 See, for comparison, Rammant-Peeters, *Pyramidions égyptiens*, pl. I–XXXIV and the overview of measures given on pp. 108–10.

cubits in total. . . . You are commanded to find out what is before it. How many men will it take to remove it in 6 hours if their minds are apt? Their desire to remove it will be small if (a break at) noon doesn't come. You shall give the troops a break to receive their cakes, in order to establish the monument in its place. One wishes to see it beautiful.

The text of papyrus Anastasi I indicates that a monument is put up with the help of a room filled with sand. In order to erect the monument, the sand needs to be transported. The information given by Hori that shall enable a scribe to calculate the workforce that is needed includes the measurements of the magazine that was filled with sand and a given time span of 6 hours. The problem is to determine how many able men it will take to remove the sand in this time. The solution would probably determine the amount of sand that needs to be transported, and then, with a given factor indicating how much sand one person can transport in a given time, the number of workers needed to complete the task could be determined. The text then points to a practical issue that was probably not part of the mathematical training, that is, in practice, a lunch break needed to be included in the estimated time.

14.3.5 Rations for an Expedition (Papyrus Anastasi I, 17,2–18,2)

The handling of rations is also indicated by Hori to have been part of Amenemope's earlier letter. Hori renders the situation as follows (papyrus Anastasi I, 6,4–6,7): Amenemope does such a bad job at it, that the official responsible for the administration of the granary refuses to put his seal on it:

> The seventh (scribe) stands receiving the rations for the soldiers. Your lists are confused and cannot be made right. Cheriuf plays the deaf man and will not listen. He takes an oath by Ptah saying: I will not let the seal be set upon the granary. He goes away in a rage.

It is, therefore, not surprising that Hori expects Amenemope to fail his problem, stated as the last problem of the mathematical section, as well. Again, Hori begins with some ironic remarks about Amenemope's capabilities:

> O scribe, keen of wit, to whom nothing whatsoever is unknown. Flame in the darkness before the soldiers, you are the light for them. You are sent on an expedition

to Phoenicia at the head of the victorious army to smite those rebels called *Nearin*. The bow-troops who are before you amount to 1900, *Sherden* 520, *Kehek* 1600, *Meshwesh* (100?), *Tehesi* 880, sum 5000 in all, not counting their officers. A complimentary gift has been brought to you and placed before you: bread, cattle, and wine. The number of men is too great for you, the provision is too small for them. Sweet *Kemeh* bread: 300, cakes: 1800, goats of various sorts: 120, wine: 30. The troops are too numerous; the provisions are underrated like this what you take from them (i.e., the inhabitants). You receive (it); it is placed in the camp. The soldiers are prepared and ready. Register it quickly, the share of every man to his hand. The Bedouins look on in secret. O sapient scribe, midday has come, the camp is hot. They say: It is time to start. Do not make the commander angry! Long is the march before us. What is it, that there is no bread at all? Our night quarters are far off. What is it, Mapu, this beating of us? Nay, but you are a clever scribe. You cease to give (us) food when only one hour of the day has passed? The scribe of the ruler—may he live, be prosperous and healthy—is lacking. Were you brought to punish us? This is not good. If Pa-Mose hears of it, he will write to degrade you.

The final section of the mathematical part of the letter broaches the issue of rations, which had been one of the central aspects of the Middle Kingdom mathematical texts. The example that is described by Hori is the provision of rations during an expedition, which includes various types of foreigners. In addition to rations that were probably carried with the expedition, complimentary gifts are received and need to be distributed. Note that once more the data given by Hori are not sufficient to actually solve the problem (e.g., the number of officers is explicitly not mentioned in the list of people given in the description of the expedition). Likewise, the factors telling how much individual groups should receive in relation to other groups (like it was indicated in the ration problems of the Rhind papyrus) are not indicated. In this example, Hori also takes the trouble to describe the immediate consequences if the scribe fails, that is, the atmosphere within the expedition will become extremely unpleasant; ultimately, failure of a scribe to fulfill the expectations will lead to his degradation.

15.

Further Aspects of Mathematics from New Kingdom Sources

Due to natural conditions, the majority of sources for our study of ancient Egypt is found in tombs and temples—cities were located then as today along the Nile and hence have often been superseded by modern cities (and are, therefore, unavailable for excavations). Moreover, the humidity found near the river is likely to have destroyed ancient evidence, especially in the form of papyri. In order to preserve papyrus, the total absence of water is needed, which explains first of all why Egypt (and its deserts) were the prime location for the preservation of ancient (not only Egyptian) papyri. This also explains the imbalance of the extant sources, originating mostly from desert sites—that is, temples and tombs. This imbalance (and a general modern fascination with mythical Egypt) has led to the presumably wrong impression that the Egyptians were constantly focused on death and afterlife, and, therefore, even mathematical practices can be found in religious texts, such as the weighing of the heart described in a section of the *Book of the Dead*.

Contrary to this idea that daily life practices were integrated into a culturally predominant religion, I would like to propose a different scenario: mathematics not only played a huge role within the education and work life of a scribe, but it also held an important role within ancient Egyptian culture in general. This role developed from the beginnings of the Egyptian state and the first uses of the number system (e.g., when the display of large numbers was used to express a certain power) and was by no means confined to the realm of the living.

15.1 MATHEMATICS AND WISDOM LITERATURE: METROLOGY IN THE *TEACHING OF AMENEMOPE*

A belief in justice ascribed to metrology has also caused the latter to appear as a topic in another genre of Egyptian literature, so-called wisdom texts or instructions. The instruction of Amenemope, usually assigned to the Ramesside Period, although the extant manuscripts are later, is extant completely in papyrus BM 10474.[1] Amenemope introduces himself as the author in the introduction:[2]

> Made by the overseer of fields, experienced in his office,
>
> The offspring of a scribe of Egypt,
>
> The overseer of grains who controls the measure,
>
> Who sets the harvest-dues for his lord,
>
> Who registers the islands of new land,
>
> In the great name of his majesty,
>
> Who records the markers on the borders of fields,
>
> Who acts for the king in his listing of taxes,
>
> Who makes the land-register of Egypt;
>
> The scribe who determines the offerings for all the gods.
>
> Who gives land-leases to the people,
>
> The overseer of grains, [provider of] foods,
>
> Who supplies the granary with grains;
>
> The truly silent in This of Ta-wer,
>
> The justified in Ipu,
>
> Who owns a tomb on the west of Senu,
>
> Who has a chapel at Abydos,
>
> Amenemope, the son of Kanakht,
>
> The justified in Ta-wer.

In this introduction, Amenemope declares himself to be involved in two kinds of measuring, that of land (responsible for the borders of fields) and of grain (who controls the

---

1   For an English translation of the entire text, see Lichtheim, *Literature II*, pp.146–63 or Simpson, *Literature*, pp. 223–43. A recent French translation with introduction and detailed notes can be found in Vernus, *Sagesses*, p.299–346.

2   Translation by Lichtheim, *Literature II*, pp. 148–49.

(grain) measure). He performs these actions for the king, but it is obvious that the ability to execute these tasks is associated with a certain power (originally coming from the king, but through the action of measuring transferred to the scribe himself). In practice, this power consisted in the control of the measuring vessel (as stated in the introduction in "who controls the measure"). Several measuring tools (and their correct usage) are then the topic of chapters 6, 16, and 17:[3]

Chapter 6

Do not move the markers on the borders of fields,
Nor shift the position of the measuring-cord.
Do not be greedy for a cubit of land,
Nor encroach on the boundaries of a widow.
[. . .]

Chapter 16

Do not move the scales nor alter the weights,
Nor diminish the fractions of the measure;
Do not desire a measure of the fields,
Nor neglect those of the treasury.
The Ape sits by the balance,
His heart is in the plummet;
Where is a god as great as Thoth,
Who invented these things and made them?
Do not make for yourself deficient weights,
They are rich in grief through the might of god.
[. . .]

Chapter 17

Beware of disguising the measure,
So as to falsify its fractions;
Do not force it to overflow,
Nor let its belly be empty.

---

3   Ibid., pp.151, 156–57.

Measure according to its true size,

Your hand clearing exactly.

Do not make a bushel of twice its size,

For then you are headed for the abyss.

[…]

In these sections, Amenemope warns strongly against the falsification of any measuring process (land measuring in progress as well as the change of the established border stones in chapter 6, weights and balances in chapter 16, and (grain) measuring vessels in chapter 17.) The punishment for actions of this kind originates from the realm of the gods; chapter 16 mentions Thot explicitly as the one who guards the balance (after having invented it in the first place).

The teachings, a literary genre with examples ranging from the Old Kingdom to the Ptolemaic Period, can also be used to trace a certain development in self-perception of the scribes over time.[4] During the Old Kingdom, Egyptian officials, who record their careers in form of autobiographies in their tombs, put great emphasis on the record of royal satisfaction with their work and the rewards and promotions they earned from the king for the successful completion of their tasks.[5] During the Old Kingdom, the success of scribes was judged according to their rank within the royal or temple administration and (linked to this feature) according to their relationship with the king. Thus, the social and cultural setting of the scribes was closely connected to the king and his needs. This can also be traced in the *Teaching of Ptahhotep*, which begins by the statement of the mayor of the city, the vizier Ptahhotep, who asks the permission of the king to train a successor ("a staff of old age"[6]) because he himself has become old (cf. the quote from *The Teaching of Ptahhotep* given in chapter 5, p. 39).

The teaching states *expressis verbis* at the beginning as well as at the end,[7] the necessity of the king's approval to appoint a successor. The maximes taught in the *Teaching of Ptahhotep* include aspects of professional life (i.e., the correct behavior toward superiors and inferiors within the administrational system, the advice to accept one's position in life, and a warning against exploiting power or acting greedily) as well as of private life (e.g., the advice to take a wife and treat her well). In the *Teaching of Amenemope*, from the New Kingdom, in

---

4 For a more detailed discussion cf. Imhausen, "Mathematik und Mathematiker."
5 See, for example, the autobiography of *Weni* translated in Lichtheim, *Literature I*, especially p. 21.
6 Ibid., p. 63.
7 "As you succeed me, sound in your body, the king content with all that was done, may you obtain (many) years of life!" Lichtheim, *Literature I*, p. 76.

comparison, the importance of metrological and mathematical knowledge of a scribe is accentuated. And it is explicitly stated that this knowledge enables the scribe to "act for the king" and to "determine the offerings for all the gods"; hence, through the metrological and mathematical knowledge a certain power, which is officially that of the king, is transferred to the scribe who performs these actions in the place of the king.

15.2 MATHEMATICS IN THE *DUTIES OF THE VIZIER*

The *Duties of the Vizier* is a text that was first discovered in the tomb of Rekhmire, who held the office of vizier (*t3jtj z3b ẞtj*),[8] the highest administrative position under the pharaoh during the time of Thutmosis III and Amenhotep II.[9] The earliest extant copy comes from the tomb of User-Amun (TT131). The text in the tomb of Rekhmire is the largest extant version. Others were found in the tomb of Amenemopet (TT29), also of the Eighteenth Dynasty, and the tomb of Paser (TT106) of the Nineteenth Dynasty.[10] Although all the sources for this composition originate from New Kingdom tombs, the office of the vizier was established much earlier, presumably as early as the Second Dynasty.[11] In the New Kingdom, the office was divided between two viziers, one for Upper Egypt and one for Lower Egypt. Rekhmire held the position of the vizier of Upper Egypt. Apart from the *Duties of the Vizier*, there are also an autobiography and the king's speech given on the occasion of the installation of Rekhmire that inform us about Rekhmire and his office.[12] From the latter, the last paragraph holds information relevant to the role of mathematics for Rekhmire:

---

8   Wolfram Grajetzki rightly reminds the modern reader to be careful with the English translation "vizier": "The word 'vizier' is of Persian origin and denotes the most important minister next to the ruler in several Islamic countries. In nineteenth-century Egyptology it was taken from there and refers in Egyptological literature to an official with the title 'tjaty.' Although they may have had responsibilities similar to those of Persian viziers, it should always be kept in mind that 'vizier' is a translation of a specific word with a specific meaning in another culture and time" (Grajetzki, *Court Officials*, p. 15).

9   First publication, Virey, *Rekhmara*. The *editio princeps* of the tomb of Rekhmire, which remains the most important source of the *Duties of the Vizier* is Davies, *Tomb of Rekh-mi-Rē*'. On the person of Rekhmire, see Dorman, "Rekhmire," and Shirley, "Viceroys, viziers & the Amun precinct." Two monographs have been published on the composition known as the *Duties of the Vizier*, the early Farina, *Le funzioni del visir* (published in 1916), and the more recent study by G.P.F. van den Boorn, which was published in 1988 (van den Boorn, *Duties of the Vizier*); see also Sethe, *Einsetzung des Viziers*. A summary of the information about the office of vizier can be found in van den Boorn, *Duties of the Vizier*, pp. 310–31; for an overview of the context of the composition, see van den Boorn, *Duties of the Vizier*, pp. 300–2.

10   Van den Boorn, *Duties of the Vizier*, p. 365. Since the study of van den Boorn, the tomb of User-Amun has been published; see Dziobek, *Gräber des Vezirs User-Amin*, and Dziobek, "Theban tombs."

11   See Grajetzki, *Court Officials*, p. 15. On viziers of the Middle Kingdom, see Grajetzki, *Court Officials*, pp. 15–41 and Allen, "High officials"; on viziers of the Old Kingdom, see Helck, *Beamtentitel*, pp. 134–42.

12   For a translation of the king's speech, see Lichtheim, *Literature II*, pp. 21–24. A translation of Rekhmire's autobiography can be found in Gardiner, "Autobiography of Rekhmerēʿ".

Furthermore, pay attention to the plowlands when they are being confirmed. If you are absent from the inspection, you shall send the chief inspectors to inspect. If anyone has made an inspection before you, you shall question him. May you act according to your charge.[13]

The vizier was the second in command (after the pharaoh) in administrative and judicial affairs. Maybe because of the exalted position that this office entailed, viziers decorated the walls of their tombs not only with depictions of themselves when receiving goods or controlling deliveries, but also with an inscription that can be used by modern readers to learn about the procedures carried out in this office and its responsibilities. Van den Boorn stated: "The text has a definite propagandistic value in presenting the vizier as the all-powerful, mighty and paramount figure at the top of bureaucracy and society."[14] The following sections from the *Duties of the Vizier* were chosen to illustrate (if only indirectly) the role that mathematics played for those that held this office.

The vizier worked closely with another official, the overseer of the treasury (*jmj-r3 ḫtm*), as is detailed in sections R6–R7 of the *Duties of the Vizier*:

Then the overseer of the treasury shall come to meet him and he shall report to him saying: 'All your affairs are sound and prosperous. Every responsible functionary (lit. 'the one belonging to the business') has reported to me saying: 'All your affairs are sound and prosperous; the palace is sound and prosperous.' Then the vizier shall report to the overseer of the treasury saying: 'All your affairs are sound and prosperous. Every department of the residence city is sound and prosperous.'[15]

This passage entails a (regular) meeting between the vizier and the overseer of the treasury, in which the overseer of the treasury confirms to the vizier that all business is as it should be, which—presumably after a check from the vizier—is then confirmed to the overseer of the treasury. One can only assume that the meeting as it is described here, entailed more than simply saying the quotes given here; for example, it could have included the presentation of some written documentation of the affairs, which would show them to be in order.

---

13   Translation from Lichtheim, *Literature II*, p. 24.
14   Van den Boorn, *Duties of the Vizier*, p. 41.
15   Ibid., p. 55.

As a modern reader might suspect from the extant official administrative documents and the variety of titles of officials, the administration in pharaonic Egypt presents itself as a sophisticated system of interrelated offices, apparently with built-in controls from the most basic level of those scribes who actually collected the revenue up to the highest instance, the vizier.[16] In an exchange between offices, like the one described here, the lower-rank party probably had to present a written account of the business that was its responsibility. The collection of revenue holds an important place within the self-perception of the viziers, as can be seen from its elaborate depiction in the tomb of Rekhmire (for parts of these registers, see figure 26).[17]

Of special importance were affairs that involved (arable) land. If there is a disagreement, the petitioner has to deal with the vizier himself (whereas in other affairs, a subordinate is sufficient to handle matters), as is indicated in R17–R19:

As for any messenger whom the vizier sends for concerning any petitioner, he (the vizier) will let him (the messenger) go to him (the petitioner). However, as for anyone who shall make a petition to the vizier concerning fields, it is to him(self, the vizier) that he orders (summons) him, in addition to consulting (lit. the hearing of) the overseer of fields and the council of the mat. He will allow him (lit. make to him) a delay of two months for his fields in Upper and Lower Egypt. But as for his fields that border on the Southern City (and) the Residence, he allows him (lit. makes to him) a delay of (only) three days, as according to the law. He hears every petitioner according to (lit. this) law which is in his hand (i.e. whose enforcement he is charged with).[18]

According to the preceding passage, it is the vizier who is ultimately charged to ensure that land measurement and the respective amounts of produce that are to be delivered, which are determined from the results of the measuring, are carried out properly. In the case of a dispute, those officials that had supposedly been present at the execution of the measuring and were therefore responsible for the correct results are called in to serve as experts in the case. Van den Boorn stresses that "the vizier apparently does not pronounce a verdict on the dispute itself. He only orders a re-assessment to be carried out by the local officials in charge and stipulates its duration. Therefore, the petitioner's request is not for a trial or re-trial, his

---

16 Some information about these offices can be drawn from Brooklyn 35.1446 papyrus from the late Middle Kingdom; see Hayes, *Papyrus Brooklyn 35.1446*.
17 See Davies, *Tomb of Rekh-mi-Rēʿ*, plates. XXIX–XXXV.
18 Translation from van den Boorn, *Duties of the Vizier*, p. 147.

FIGURE 26: Detail of the collection of taxes from southern towns of the tomb of Rekhmire (drawing by Nadine Eikelschulte).

appeal is concerned with obtaining legal permission for a re-assessment by the local authorities."[19] Further sections also deal with the vizier and land measurement, as, for example, R26:

It is he (the vizier) who dispatches the group (of) scribes of the mat to execute the instruction(s) of the Lord. It is for the matters that are heard concerning any field, that there has to be the document of the (pertinent) town district (i.e., the local land registry) in his office. It is he who establishes the boundaries of every domain, (of) every vegetable plot, (of) every divine offering, (of) everything to be sealed (officially in this respect). It is he who enforces every promulgation (of the king).[20]

The preceding section deals with land measurement, probably in rural areas, to which the scribes of the mat are sent to carry out the measuring.[21] While they are the ones who carry out the actual procedures that establish boundaries and the like, the power to transform the results of the measuring into valid facts lies with the vizier, who does so for the king. In order to ensure a correct procedure that can be verified in case of a dispute, the importance of written documentation is especially stressed.

15.3 MATHEMATICS AND DEATH

Like the *Duties of the Vizier* of the previous section, the evidence in this section comes from a funerary context. While it is probably more a result of the vagaries of preservation (or in this case, not really vagaries but the direct consequence of the tombs being located in the desert, where perfect conditions for the preservation of artifacts were at hand) than a reflection of an Egyptian focus on the afterlife that we have so many sources from a funerary context (and so little from daily life, which happened to be close to the Nile, i.e., close to water, which is advantageous for life but the contrary for the preservation of artifacts). Nevertheless, the ancient Egyptians certainly did make some effort to ensure their well-being after death, and, as mentioned before, due to the conditions of their tombs, many artifacts have survived. It is interesting to find among these artifacts even some "mathematical" objects— clearly an indication of the value and importance that was attributed to mathematics and its applications.

---

19   Ibid., p. 165.
20   Ibid., pp. 265–66.
21   Ibid., pp. 266–67.

A prominent example can be found in the judgment of the dead. After death, the deceased was believed to have to undergo a trial, during which his life on Earth was judged according to the *Maat*, the Egyptian concept of morally correct conduct.[22] Should the deceased fail this trial, that is, if he had lived in a way that was not in accordance with *Maat*, he would be devoured by Ammit, a monster whose body was part crocodile, part lion, and part hippopotamus (i.e., composed of those animals that were known to be dangerous to humans). Should he pass, the way to eternal life in the hereafter was open to him. The earliest representation of the judgment scene in the Theban necropolis is found in the tomb of Menna (TT69).[23] This scene is also depicted in a vignette of the *Book of the Dead*, a New Kingdom composition, where it is often associated with spell 125.[24] The vignette depicts the judgment hall, where the deceased is shown with several gods and goddesses next to a scale. Although the individual figures who are present in this scene vary over time, the gods Thoth (who records the result of the judgment in writing) and Osiris (as presiding chief judge) are usually represented as is Ammit the devourer.

The trial consists of two parts, the "negative confession" (or declaration of innocence) of the deceased, who addresses the forty-two divine judges individually and denies that he ever committed a certain set of specific sins. This is the content of spell 125 of the *Book of the Dead*, the beginning of which reads:[25]

Spell for descending to the broad hall of the Two Truths.

Hail (to thee), great God, lord of the Two Truths. I have come unto thee, my Lord, that thou mayest bring me to see thy beauty. I know thee, (I know thy name,) I know the names of the 42 Gods who exist with thee in this broad hall of the Two Truths, who live on supporters of evil and sip of their blood on this day of taking

---

22 On the judgment of the dead, see the overview in Stadler, "Judgment after death," and the monograph Seeber, *Totengericht*.

23 For descriptions of this tomb, see Hartwig, "Tomb of Menna," and Hodel-Hoenes, *Life and Death*, pp. 85–111. The judgment can also be found in the decoration of the Hathor temple at Deir el Medina; see most recently von Lieven, "Book of the Dead, Book of the Living," esp. pp. 263–64, with references to earlier literature.

24 The title *Book of the Dead* is the modern name for compositions consisting of a collection of spells (or chapters) that were supposed to provide protection for the dead and support his passage to the afterlife. The Egyptian title was *Spells for Going Forth by Day*. There are about 200 of these spells; the origin of some can be traced back as far as the *Pyramid Texts*. For a detailed and richly illustrated account of the ancient Egyptian attitudes to death, see Taylor, *Death and the Afterlife*. For an English translation of the *Book of the Dead*, see, for example, Allen, *Book of the Dead*. Apart from spell 125, the vignette showing the judgment is also combined with spell 30B, in which the heart is asked not to speak against his owner.

25 Translation from Allen, *Book of the Dead*, p. 97.

FIGURE 27: Weighing of the heart from the *Book of the Dead* of Hunefer

account of characters in the presence of Unnofer. Behold, I am come unto thee. I have brought thee truth; I have done away with sin for thee.

I have not sinned against anyone. I have not mistreated people. I have not done evil instead of righteousness. I know not what is not (proper); I have not done anything bad. I have not at the beginning of each day set tasks harder than I had set (previously). . . .

The Egyptian expression for "this day of taking account of characters" from the preceding quote is *hrw pwy n ḥsb qdw*, using the verb *ḥsb* well known from the mathematical texts as "calculate," especially in an administrative context. This leads to the second part of the judgment, which complements the spell, the procedure of weighing the heart of the deceased against the goddess Maat, either represented as a depiction of a female with a feather on her head or only by a feather (indicating that the heart to be weighed against her is supposed to be light and unburdened by any guilt).[26] The weighing of the heart from the vignette from the *Book of the Dead* of Hunefer is shown in figure 27.[27] In the vignette, from which the weighing scene of figure 27 is taken, some persons appear several times (Hunefer is shown three times and the god Anubis, twice), which indicates that several successive stages of the judgment are combined in one illustration (like a graphic novel put into a single image), without any

26   The combination of the judgment of the dead with the weighing of his heart is first attested on a stela of the Eleventh Dynasty; see Seeber, *Totengericht*, p. 1.

27   For the complete vignette, see http://www.britishmuseum.org/explore/highlights/highlight_objects/aes /p/page_from_the_book_of_the_dead.aspx (July 8, 2014).

indication of the borders between them. The top row shows Hunefer (who was a scribe under Seti I. of the Nineteenth Dynasty) adoring a row of fourteen gods, which presumably represent part of the forty-two gods to whom the individual negative confessions are given. The scenes underneath are to be read from left to right (despite the text in cursive hieroglyphs, which reads from right to left). They begin with Hunefer being led to the scale by Anubis. Anubis is then shown in the next sequence to supervise the measuring of the heart of Hunefer. The heart is depicted on the left tray of the scale as it is weighed against the feather representing Maat, which is placed on the right tray. The scale used in the judgment of the dead depictions is a balance scale shown to be approximately as tall as a person.[28] In front of the scale, Ammit is ready, should Hunefer fail the judgment. The fulcrum of the scale is furnished with a depiction of the goddess Maat, thus divinely ensuring that the weighing process will be executed correctly. To the right of the scale, the god Thoth, who fixes the outcome of the weighing process in writing, is depicted. Thus, the Egyptian obsession of measuring and recording is performed a last time, before the deceased Hunefer can progress to his final destination. Having passed the judgment, Horus leads Hunefer to Osiris, the god presiding over the netherworld, seated on a throne at the right. Behind Osiris, the goddesses Isis and Nephtys are shown.

Apart from the scale, which is used prominently in the weighing of the heart, another measuring device seems to have had a religious role, the so-called cubit-rods. Cubit-rods are staffs that indicate the length of one cubit and may also include its subdivisions.[29] Relatively few Egyptian cubit-rods are preserved.[30] Dieter Arnold gives the following short and succinct description of cubit-rods:[31]

> They resemble our yardsticks, being short (52.5 centimeters) and stout, with a quadrangular section and a fifth side produced by chamfering the front top edge. The front, the oblique, and the top faces were used for marking and numbering the single measurements. The rear and undersurface were often inscribed with the name of the owner. The measurements indicated consist of the 7 palms and digits and their subdivisions. Their lengths vary from 52.3 to 52.9 centimeters, reminding us that ancient measures were not so standardized as those of today and that

28 For a detailed description of this scale, see Seeber, *Totengericht*, pp. 67–70.
29 The earliest studies are Lepsius, *Alt-aegyptische Elle*, and several sections in publications of William M.F. Petrie; for references see Arnold, *Building*, p. 283, note 1. See also the article Scott, "Egyptian Cubit Rods."
30 Lepsius, *Alt-aegyptische Elle*, lists about fourteen more or less complete cubit-rods, and since then not many more have been found.
31 Arnold, *Building*, p. 251.

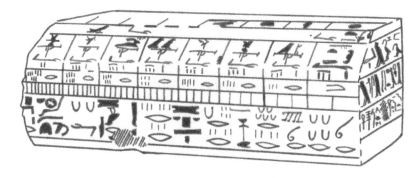

FIGURE 28: Cubit-rod of Amenhotep (Eighteenth Dynasty) (after Schlott-Schwab, *Ausmasse*, plate X)

such discrepancies have to be taken into account in our calculations of Egyptian buildings.

The shape of the cubit-rod, which Dieter Arnold described, can be seen from the drawing of the fragment of the cubit-rod in figure 28.

Apart from the (often rough) wooden cubit-rods, which may have been used in daily affairs of measuring, there are several cubit-rods made of stone and beautifully inscribed in great detail, not only with the basic subdivisions of the cubit in palms and fingers, but also with other texts relating metrological systems with cultic and other information. It is generally assumed that these specimens were not used for everyday measurements but served in a cultic context in either a tomb or temple.[32]

Adelheid Schlott-Schwab distinguishes three types of cubit-rods: (1) those actually used for measuring, which are usually made of wood (extant examples date from the Twelfth Dynasty to the Late Period), (2) those that are inscribed with offering formulas and the name of their owner, often made from stone (but there is also one example of a gold cubit measure[33]), which were part of the burial goods of high officials (extant examples date from the Eighteenth Dynasty to the Ptolemaic Period), and (3) cubit-rods that usually do not contain offering formulas but instead texts relating to the geography of Egypt and other texts, which were also made from stone and possibly kept in temples.[34]

---

32 For images of these votive cubits, see Clagett, *Egyptian Mathematics*, figures IV.24–27.
33 For a discussion of the respective burial, see Meskell, "Intimate archaeologies."
34 Schlott-Schwab, *Ausmasse*, pp. 53–63.

Both of the latter two religious uses of an original everyday metrological instrument demonstrate the importance attached to metrology and, in particular, to its instruments. The basic object for ensuring the correct measuring of length is imbued with further roles of providing guarantees, be it the warranty for accurate measurement in the netherworld (cubit-rods of high officials inscribed with offering formulas) or the guarantor for the safety and accuracy of other information, as in the case of the cubit-rods found in temples.

15.4 MATHEMATICS IN ARCHITECTURE AND ART

Among the most tantalizing subjects for historians of mathematics and, likewise, Egyptologists is the study of mathematics derived from ancient Egyptian architecture. The impressive monuments left from all periods of Egyptian history unfortunately tell us very little about the previous planning or necessarily about the construction techniques. As a consequence, ancient Egyptian monuments have been the object of all kinds of speculations. The same trust that was put in modern mathematical terminology and concepts when analyzing Egyptian mathematics was put into modern architectural plans in the analysis of Egyptian architecture, with similar "results."[35] Corinna Rossi has given a very detailed and insightful account of the available evidence, possible conclusions to draw from it, as well as a critical discussion of some theories that have become truths and were in serious need of reassessment in light of the actual evidence.[36] Despite our limited concrete evidence, it cannot be denied that mathematics was an integral part of Egyptian architecture. Consequently, architectural evidence can be used to reveal further insights into mathematical concepts. The following section presents some examples of such evidence.

A small number of architectural sketches from various times of Egyptian history are still extant.[37] However, it is only in exceptional cases that they can be assigned to remains of existing structures, as in the examples of the drawings of papyrus Turin 1885 (tomb of Ramses IV) and the Ostracon Cairo CG 25184 (tomb of Ramses IX). In only one case, an architectural sketch was found within the confines of the site it depicts.[38] Simply measuring the drawings and looking for matches in architecture is bound to fail, because ancient Egyptian drawings

---

35   Compare, for example, the plans of the temple of Sesostris I. at Tôd by Badawy and Arnold—and the actual archaeological remains (Rossi, *Architecture and Mathematics*, pp. 46–47, figures 30 and 31).

36   See Rossi, *Architecture and Mathematics*. This book, together with Arnold, *Building*, presents a comprehensive introduction to Egyptian architecture.

37   For an overview of available sources, see Rossi, *Architecture and Mathematics*, pp. 104 and 113; for a discussion of various types of drawings and their uses, see Arnold, *Building*, pp. 7–9. See also Heisel, *Antike Bauzeichnungen*, pp. 76–153.

38   For a discussion of this object, see Polz, "An architect's sketch."

FIGURE 29: Ostracon BM 41228: An architect's sketch

were executed following specific conventions, which are—as should be expected—not those used in our modern technical drawings. First of all, similar to the drawings in the earlier mathematical texts, Egyptian architectural drawings are not to scale. Instead, they use annotations to clarify their meaning. Hence, it is futile to attempt to interpret too much from a drawing alone. Any annotations given in them must be read carefully. And even if the drawing is apparently straightforward and includes ample annotations, there still may be more than one possible way to read it.

A good example of the difficulties in interpreting such a drawing is the Ostracon BM 41228 (see figure 29), which was found in the mound overlying the Eleventh Dynasty temple of Mentuhotep II at Deir el Bahri. By paleographical means, it was dated to the Eighteenth or Nineteenth Dynasty, which would be in line with the date of other finds from the same area. It was published by Stephen R. K. Glanville in 1930.[39]

The information given on the ostracon includes (besides the pictorial elements) the words for length (*3w*) and width (*wsḫt*), several numbers (27, 14, and 6, with one indication that those are to be understood as cubits), and a further line of explanatory text, which is only partly preserved: "Who is in front of it, its west rests . . . ." Glanville interpreted the plan and its annotations as follows:[40]

---

39   Glanville, "Working plan of a shrine."
40   Ibid., pp. 237–38.

The outer rectangle presents no difficulties, since the two main measurements are clearly indicated—'breadth 27', 'length 27'; but the thickness of the wall may only be guessed and the precise shape of the doorways is not indicated by their representation in elevation. It is not even certain whether the two pairs of vertical lines which are cut short by the broken top edge of the ostracon indicate merely a second doorway or a passage leading to another part of the building as well. I can make nothing of the marks immediately south of this doorway: that on the right is apparently part of the plan; the other is perhaps a sign. So far the plan shows a building 27 cubits square with the gateways in the centre of its north and south walls, and possibly a passage extending northwards from the former. The problem is to interpret the plan of the smaller building enclosed by the main square. The smaller building is also rectangular; it also has two doorways in its north and south sides respectively and bridging the N.–S. axis as in the case of the outer construction. The measurements given are 'breadth 6', 'length 14', and by analogy with those of the larger rectangle should give us the two main measurements over all.

Glanville himself remarked upon the disparity of the drawing on the ostracon and a modern plan drawn after his interpretation, concluding that this was due to the lack of scale in the Egyptian sketch.[41]

Another interpretation was offered in 1986 by Charles C. van Siclen:[42]

The pillared building whose plan is depicted on the ostracon is shown as a structure with inner and outer parts. The length and width of each of these parts is shown. The position of the doorways and the general presence of pillars are indicated. The walls of each part are given in single line, and they have no true thickness on the plan. As such, the pillars adjacent to the wall line need not be free-standing (as understood by Glanville) but equally may be part of the wall, or both. From the inscription below, it seems that the structure is aligned on a north-south axis. . . . Working with the information from the ostracon, and assuming that there was in use a wall and pillar thickness module of 1 and 4/7 cubits (the module '1 and 4/7 cubits' or '1 cubit, 4 palms' has been selected as typical. . . .), it is possible to reconstruct the plan of a typical building type: a peripheral bark shrine. The key to this reconstruction is the

41 Ibid., p. 238.
42 Van Sicklen, "Ostracon BM 41228," pp. 71–75. For a discussion see Rossi, *Architecture and mathematics*, p. 105; see also Polz, "An architects sketch", pp. 238–39.

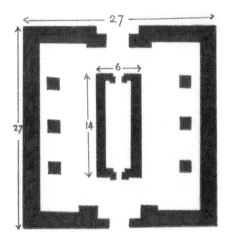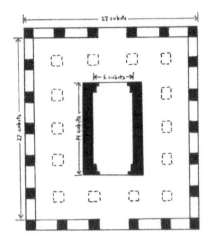

FIGURE 30: Reconstructions of the structure represented on BM 41228 by Glanville (left) and van Siclen (right). (Drawn after Glanville, "Working plan of a shrine," figure 1; and van Siclen, "Ostracon BM 41228," figure 3.)

hypothesis that for each pair of dimensions given, one of the figures represents an external dimension; and the other, an internal dimension. As I understand them, the four dimensions of the ostracon are:

external width of the outer structure—27 cubits
internal length of the outer structure—27 cubits
internal width of the inner structure—6 cubits
external length of the inner structure —14 cubits

To arrive at the correct overall external dimensions for the various parts of the building simply requires the addition of the wall thicknesses, that is, twice the module, to the internal figures. (Missing internal dimensions are achieved by subtracting twice the module from the given exterior figures.) The outer structure thus becomes 30 and 1/7 by 27 cubits in external dimensions; and the inner one, 9 and 1/7 by 14 cubits. If the inner structure is centered, the surrounding interior space is 6 and 1/2 to about 7 and 1/3 cubits on a side. Assuming that the pillars were evenly spaced about the sides of the structure—roughly 3 cubits apart, there would have been six pillars along the front and rear of the building and seven along each side. The resultant plan of a peripteral temple shows what I believe to have been the intentions of the ancient architect as derived from the ostracon.

Remains of actual structures similar to the suggested reconstructions (as shown in figure 30) can be found for both cases, and at this point it seems not possible to decide which of them is the one intended by the ancient architect. The two interpretations make use of different presuppositions—while Glanville's reconstruction attempts to take the drawing and the annotations at face value, van Siclen assumes certain conventions that allow him to interpret the drawing more freely. Ultimately, it is only the use of drawings of which we also have securely identified archaeological remains that will allow us to learn about the conventions used in such objects.

Evidence of reliefs and paintings also allow access to an Egyptian conceptualization of space. Egyptian depictions of objects of all kinds have been characterized as following one guiding principle, that is, the composition of typical aspects of the object.[43] This principle is also evident in the sketches of the hieratic mathematical texts often found with geometric problems.[44] They represent a typical shape of the respective object, which may differ from the actual proportions of the specific case.

Established conventions included the combination of several views in a meaningful way, as can be demonstrated by the example of the depiction of a lake surrounded by trees (figure 31). Thus the paintings and reliefs found in tombs contain "encoded" information similar to the architectural sketches of the previous section. The order and proportion of objects within Egyptian art follow rules that can be deducted from the available source material in form of reliefs and paintings.[45] Spatial relations are expressed using specific conventions, such as the overlapping of items to indicate their relationship in reality. Depth may also be indicated through a "pile," as often seen in the depiction of offering tables, but also in the representation of humans, as, for example, in a group of prostrate figures (figure 32) in the tomb of Nebamun (BM EA 37978) from the Nineteenth Dynasty.[46]

In order to achieve required compositions, some artists made use of square grids.[47] These were applied on the prepared surface using rulers or (more often) a length of string dipped into red paint, which was stretched across the surface and snapped against the wall.[48] Within this more or less precise grid, preliminary sketches of the required scenes were drawn, also in

---

43 Robins, *Proportion*, pp. 3 and 13.
44 This is most striking when it comes to the representation of three dimensional objects, as in the case of the pyramid problems (Rhind papyrus, numbers 56–60) or the problem of a truncated pyramid (Moscow papyrus, number 14).
45 For a detailed study see Robins, *Proportion*.
46 Parkinson, *Nebamun*, p. 95.
47 The use of square grids in the design of reliefs and paintings is attested from the Middle Kingdom on, see Robins, *Proportion*, p. 26. For a well-illustrated discussion of the individual stages of composition see Robins, *Proportion*, pp. 23–26. Rossi, *Architecture and Mathematics*, p. 122, considers the use of square grids for architectural plans on papyrus unlikely, which is borne out by the surviving evidence.
48 Robins, *Proportion*, p. 26.

FIGURE 31: Depiction of lake with surrounding trees from the tomb of Nebamun

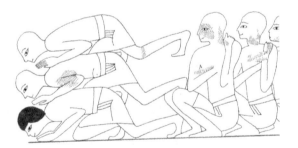

FIGURE 32: Depiction of farmers from the tomb of Nebamun

red. The compositions had to be approved before the final stage, the creation of the scene in paint or relief, was begun. The measurements of these grids have been analyzed by Eric Iversen and Gay Robins.[49] Although there seem to be preferred lengths, they should not be seen as a basis for a metrological unit.[50]

---

49 Iversen, *Canon and Proportions*, and Robins, *Proportion*.
50 Explicitly on this question see Robins, "Canonical proportions and metrology."

In his analysis of Egyptian descriptions of statues, Friedhelm Hoffmann has established several aspects of Egyptian concepts of spatial relations used in art.[51] First of all, the term for designating height, *q3*, often, but not always, corresponds to our understanding of the "height" of a statue. The determining factor of *q3* is its relation to the line of sight of the viewer. Thus, *q3* designates a length that is perpendicular to the line of sight. Usually, this will coincide with our idea of height, unless, as in one example cited by Hoffmann, the statue is still lying in the quarry. Furthermore, various elements of the statue, such as its base, are included in its given height. An analysis of metrological aspects of ancient objects must take account of these ancient conventions.

---

51   Hoffmann, "Measuring Egyptian statues."

16.

Summary

Despite the scarcity of mathematical texts, the New Kingdom offers a variety of sources that allow us insights into the uses and roles of mathematics at that time. As in previous periods, mathematics and the scribes who practiced it expertly continue to play a vital role in the administration. In order to illustrate the type and style of available sources, extracts of Papyrus Harris I and the Wilbour Papyrus have been used. Both of these are as interesting as they are complex in terms of their usage of mathematical techniques and concepts. However, apart from these two outstanding papyri, it is the large number of smaller accounts written on papyri and ostraca that clearly documents the ongoing significance of administrative mathematics during the New Kingdom. Like Lahun, the town of the Middle Kingdom, yielded the largest papyrus finds for that period, so did the town of Deir el Medina for the New Kingdom, providing unique insights into the daily life of an ancient Egyptian community of that time.[1]

By the time of the New Kingdom, the importance of the mathematical work of the scribes for the king has become part of their self-conception and therefore appears not only in the products of their daily work (the accounts and other administrative documents), but also in the literary texts. Two literary compositions make frequent references to mathematics and metrology, on the one hand, the *Late Egyptian Miscellanies*, and on the other hand, the *Teaching of Amenemope*. While the *Miscellanies* stress the amenities that come with the scribal profession and the power that a scribe holds who performs metrological duties, the *Teaching of Amenemope* links the metrological duties to the execution of justice and the realisation of a life according to the *Maat*.

---

1 A variety of websites are available to help access the material of Deir el Medina, see e.g. The Deir el Medina Database at www.leidenuniv.nl/nino/dmd/dmd.html (accessed July 13th, 2015).

This connection of mathematics and justice is also apparent in ritual and religious sources. Mathematical or metrological practices are perceived as a guarantor for justice, hence the use of a scale in the judgement of the dead. While oral confessions may be given falsely, the weighing of the heart against Maat supposedly enabled an inerrant judgement about the character of the deceased.

GRECO-ROMAN PERIODS

The last part of the history of ancient Egypt can be characterized as a combination of traditional Egyptian culture and various foreign influences.[1] During the first millennium BCE, Egypt fell under foreign rule several times, most notably during the First Persian Period (525–404 BCE), which lasted over 100 years.[2] However, even during Egyptian rule, intense foreign relationships can be documented, such as in the famous Amarna letters, the correspondence between the Egyptian court and kings of Assyria, Babylon and the Mitanni and Hittite empires, among others.[3] Most of the 350 letters written in cuneiform script on clay tablets were written in Babylonian, the *lingua franca* in interregional diplomacy and trade.

The Mesopotamian dominance was confirmed militarily by Esarhaddon (680–669 BCE) and Ashurbanipal (668–ca. 630 BCE), when they conquered—albeit for a brief time—much of the ancient Near East, including Egypt.[4] It is probable that during this time some exchange of Egyptian and Mesopotamian knowledge (including that of mathematics) took place. The Egyptian texts from the Greco-Roman Periods show several indicators that contact between the Mesopotamian and Egyptian mathematical culture had indeed taken place, or at least that Egyptian mathematics and astronomy were influenced by its Mesopotamian neighbor.[5]

---

1. For a brief overview of Egypt during this time see Ellis, *Greco-Roman Egypt*, Lloyd, "Ptolemaic Period," and Peacock, "Roman Period." For a more detailed introduction, see Bowman, *Egypt after the Pharaohs*, and Chauveau, *Egypt in the Age of Cleopatra*.
2. For an overview of the Late Period, see Lloyd, "Late Period." Egypt's relationship to other countries is discussed, for example, in Betlyon, "Egypt and Phoenicia in the Persian Period" and Holladay, "Judaeans (and Phoenicians) in Egypt."
3. For an English edition, see Moran, *Amarna Letters*.
4. Kuhrt, *Ancient Near East*, p. 499. For Ashurbanipal's campaigns against Egypt, see Spalinger, "Assurbanipal and Egypt."
5. This has been stated since the beginning of research of Demotic mathematical and astronomical papyri; see Parker, *Demotic Mathematical Papyri*, p. 6, and also Høyrup, *Lengths, Widths, Surfaces*, pp. 405–6. However, a detailed study of this influence or exchange has as yet to be done.

Another foreign presence can be documented again at least from this time on—that is, that of Greek traders. According to Diodorus Siculus, Psammetichus I (664–610 BCE) encouraged trade from Greece and Phoenicia; Greek trading stations, of which Naucratis was the most famous, were established by 625 BCE.[6]

In 332 BCE, Alexander the Great conquered Egypt and founded Alexandria before moving on. After his death in Babylon in 323 BCE, a succession crisis erupted, in which Ptolemy was appointed to be satrap of Egypt. He then established himself as ruler in his own right and took the title of king in 305. He founded the Ptolemaic Dynasty, which ruled Egypt for nearly 300 years. Egyptian temples continued to have a privileged status under the Ptolemies, who took on the Pharaonic duty of their construction and renovation. After the conquest of Egypt by Octavian (Augustus) in 30 BCE, Egypt became a province of the Roman Empire. The Emperor Augustus assumed the position of pharaoh. He continued the previous system of administration used by the Ptolemies.

The use of mathematics in daily life is apparent in various documents. Land and its usage remained a central issue. Scribes recording results of land surveys were based at temples, which functioned as administrative centers.[7] Although the related documents do not provide us with specific mathematical procedures, their content indicates that calculations of some kind were behind the numbers that appear in the texts.[8] Numerous land leases have been published by Heinz Felber.[9] Within their formulaic structure, stipulations concerning the rent and harvest tax are made, which consist of amounts of grain, either indicated as a fraction of the harvested grain or indicated as a fixed amount. Tax receipts constitute another large group of sources for documents related to mathematical practice in this time.[10]

Egyptian sources from the Greco-Roman Periods are written in demotic, a cursive script, which began to develop from the mid-seventh century BCE. Like its hieratic predecessors, it was written exclusively from right to left. Where individual signs can be distinguished even within ligatures in the hieratic script, in demotic, whole groups of signs have become conflated into one ligature, and sometimes the original underlying hieroglyphic signs can no longer be determined. Demotic designates not only a new stage in the script of ancient Egypt, but also the penultimate stage in the Egyptian language, to be followed by Coptic. Greek and Egyptian cultures did not only coexist, but they also influenced each other, as is seen by translations of

---

6   Kuhrt, *Ancient Near East*, p. 641. See Möller, *Naukratis*, for a detailed description of Naucratis.
7   See Manning, *Land and Power*, p. 147.
8   For an overview of the use of land in Demotic Egypt see Felber, *Demotische Ackerpachtverträge*, and Manning, *Land and Power*.
9   Felber, *Demotische Ackerpachtverträge*.
10   See, for example, Muhs, *Tax Receipts*.

literature. At the moment at least, it seems that more was transferred from Egypt to Greece than vice versa—which would be in line with the credit Greek historians give to Egypt.[11] Despite the evidence of references to traditional Egyptian literature, however, the demotic literature has to be seen in its own right and not simply as a continuation of a previous time.[12] The same is true for the second corpus of mathematical texts extant from ancient Egypt. They are written in demotic and date to the Greco-Roman Periods. While influences from Mesopotamia and traditions from earlier Egyptian mathematics can be documented, there are also new aspects that are a distinct part of the character of demotic mathematics, as I hope to illustrate in the following chapter.

---

11   Quack, *Altägyptische Literaturgeschichte III*, p. 172.
12   For an overview and examples of demotic literature, see Quack, *Altägyptische Literaturgeschichte III*, and Tait, "Demotic literature." English translations of demotic texts can be found in Simpson, *Literature*, pp. 443–529, and Lichtheim, *Literature III*, pp.125–217. See also the collection of German translations in Hoffmann and Quack, *Anthologie der demotischen Literatur*.

17.

Mathematical Texts (II): Tradition, Transmission, Development

17.1 OVERVIEW OF EXTANT DEMOTIC MATHEMATICAL PAPYRI

The Greco-Roman Period is the second time in Egyptian history for which a significant corpus of mathematical texts happens to be extant.[1] They are written in the Egyptian script and language of that time, that is, demotic, and are, therefore, commonly known as "demotic mathematical texts." Around 1500 years are between the hieratic and demotic mathematical texts, and some significant changes in Egyptian mathematics happened, as will be outlined in the later part of this chapter. It should be noted, however, that this is mainly an overview of the extant sources and some questions that can be raised—as our knowledge of the Egyptian culture during that period is still growing rapidly at the moment. It should also be noted that the material that is available from this period is immensely rich and would necessitate writing a second book on mathematics in Egypt during the Greco-Roman Periods, which would then include the mathematical texts (and the contemporary Greek and Seleucid material), the relevant administrative material, as well as other related texts (e.g., the inscription about fields on the walls of the Edfu temple). For this book, however, the demotic mathematical papyri constitute

---

1   Translations of the demotic mathematical problems are taken from the respective editions of Parker. These are now more than 25 years old and in need of a reassessment, which I hope to provide in the near future. However, for the time being I have chosen to use Parker's work for the sake of consistency.

the final indigenous mathematical texts available from ancient Egypt; therefore, this section was included as is despite its obvious potential to be further developed.

17.1.1 Papyrus Cairo JE 89127-30, 89137-43

The largest extant demotic mathematical text can be found on the verso of the *Codex Hermopolis*, a demotic legal text. It was discovered at Tuna el-Gebel (Hermopolis West) in 1938–39 and is today kept in the Egyptian Museum in Cairo.[2] Its mathematical part was published in 1972 by Richard Parker.[3] The text consists of eleven pieces, which in total make up a papyrus of more than 2 m in length and 35 cm in height. It contains 40 problems, some of which are too fragmentary to yield much information. On the basis of paleography, Parker dated it to the third century BCE.[4] There are several groups of problems of the same kind; for example, numbers 7–12 deal with the area of a piece of cloth, numbers 24–31 are examples of the "famous" pole-against-the-wall problem used to prove a Mesopotamian influence in demotic mathematics, and numbers 36–38 deal with geometrical figures inscribed in circles.

17.1.2 Papyrus BM 10399

The second-largest demotic mathematical papyrus was bought by the British Museum in 1868 as part of the Robert Hay collection. Its original provenance is unknown. The text published by Parker consists of $1\frac{1}{3}$ columns of a papyrus of an original height of 36.5 cm and two fragments of a third column. The verso contains $1\frac{2}{3}$ (well-preserved) columns and fragments of three further columns. This published section contains twelve problems, among which are four problems concerning the volume of a cone-shaped mast and six problems concerning fraction reckoning. Based on its paleography, it was dated by Parker to be Ptolemaic and later than the Cairo mathematical papyrus. Further parts of this papyrus were found and described in 1994 by Carol Andrews.[5] While studying the papyrus in 2006, I found that yet more of it had come to light through a papyrological exercise.

---

2   Parker, *Demotic Mathematical Papyri*, p. 1. The Codex Hermopolis is the largest known demotic legal text. It was published by G. Mattha and G. R. Hughes in 1975 (Mattha, *Demotic Legal Code*), and another translation was published in 1990 (Donker van Heel, *Legal Manual of Hermopolis*). For a brief summary of its content and publications, see Lippert, *Demotisches juristisches Lehrbuch*, pp. 153–59. Based on paleography and content, Codex Hermopolis was copied in the third century BCE, presumably from an early demotic text (Lippert, *Demotisches juristisches Lehrbuch*, pp. 155–58).
3   Parker, *Demotic Mathematical Papyri*, pp. 13–53 and pl. 1-14.
4   Ibid., p.1.
5   Andrews, "Unpublished demotic papyri."

17.1.3 Papyrus BM 10520

Only part of this papyrus has mathematical content; apart from twelve mathematical problems and one table written in seven columns on the recto, it contains parts of a land register, lists of names, and accounts. Provenance of the papyrus is unknown. According to the museum records, Memphis is suggested as the possible origin. At the bottom of column A, a date is noted, indicating the twenty-first day of *šmw* of year 5, without mentioning the name of a ruler. Museum records date the papyrus to the early Roman period. Out of more than 7 m of length, the mathematical part comprises only 89 cm. The height of the text is 25 cm. The individual mathematical problems are separated by long horizontal lines. They were published by Parker in 1972.[6] Most of them are arithmetical in nature, including multiplications, fraction reckoning, and the extraction of square roots. The last two problems, however, are geometrical and deal with the calculation of the area of rectangular fields.

17.1.4 Papyrus BM 10794

Again, provenance of this fragment is not known. It measures 13 by 5 cm and contains two multiplication tables for fractions. It was first published in Parker 1972.

17.1.5 Papyrus Carlsberg 30

This document consists of two fragments, which together measure 31 by 6.5 cm. It originated from Tebtunis and was dated by Parker to the second century CE. It contains parts of five problems; however, due to their fragmentary nature not much more can be determined.

17.1.6 Papyrus Griffith I E7

This fragment, now held in the Griffith Institute, Oxford, was bought by Grenfell in 1898 at Dimê, Soknopaiou Nesos.[7] Parker, who published its five problems in 1959, considered its height of 32.4 cm close to its original height and established a maximum width of 13.3 cm.[8] Based on the paleography, Parker dated the fragment to the end of the Ptolemaic or the beginning of the Roman Period. Only the recto is of mathematical content; the verso contains

---

6   Parker, *Demotic Mathematical Papyri*, pp. 53–63 and plates 15–18.
7   Ibid., p. 275.
8   Ibid.

another nonmathematical text.[9] The fragment contains two-thirds of a well-preserved column and enough of a second column to ascertain its mathematical content. The formulaic character of the text enabled Parker to restore the text of the individual problems. All five problems are of the same type; an unknown quantity is modified and the result of this operation is given—that is, these are successors of the hieratic ⊂ḥ⊂ problems.

17.1.7 Papyrus Heidelberg 663

Three fragments (4.3 by 19.5 cm, 6.5 by 18.3 cm, and one smallish fragment) of unknown origin exist, now kept at Heidelberg and published by Parker in 1975.[10] They contain remains of two columns in which four geometrical problems, each text supplied with a sketch, were written. The drawings show trapezoid areas, which are subdivided into smaller areas.

In addition to these papyri, several ostraca of mathematical content have been discovered, and it is likely that more exist in museums all over the world.[11]

17.2 DEMOTIC ARITHMETIC

17.2.1 Multiplication

At least some of the hieratic mathematical papyri provide us not only with procedures to solve mathematical problems, but also with some written calculations (i.e., multiplications and divisions), which enable us to follow Egyptian mathematics down to the level of arithmetic. The demotic mathematical papyri include specific arithmetical problems concerning multiplications and divisions, thereby providing evidence to contrast arithmetical techniques of both periods. An example can be found in the third problem of the papyrus BM 10520.

PAPYRUS BM 10520, NUMBER 3[12]

Calculate 13 times 17.

You shall calculate 10 times 10 as 100.

You shall calculate 3 times 10 as 30.

You shall calculate 7 times 10 as 70.

---

9   Ibid., p. 275.
10   Parker, "Mathematical exercise."
11   Ritter, "Egyptian mathematics," p. 134; note 27 lists eight examples.
12   Parker, *Demotic Mathematical Papyri*, p. 65 (no. 55). The translation given here is based on that of Parker but was modified, rendering the verb *jrj* as in the hieratic texts as "calculate" instead of "reckon."

You shall calculate 3 times 7 as 21.

You shall add the amount (of) its being: 221.

You shall say: 221.

Similar to the hieratic technique, the multiplication is split up into multiplications that can be carried out mentally. In this example, the scribe chose to calculate 10×10, 3×10, 7×10, and 3×7, all of which are within the limit of a ten times table, which one could imagine to have been learned by heart. After the individual multiplications, their results are added—supposedly, again, mentally or using an external auxiliary—and the result of the multiplication is announced.

One might wonder about the existence of multiplication tables at this point. While there are no multiplication tables preserved within the hieratic mathematical corpus and the well-documented method of carrying out multiplications in written form renders their existence at that point unlikely, there is a multiplication "table" of sorts within our demotic mathematical corpus, just before the multiplication problem presented here. It reads as follows:

PAPYRUS BM 10520, NUMBER 2[13]

64

128

192

256

320

384

448

512

576

640

704

768

832

896

960

1024

13   Problem 54 in Parker, *Demotic Mathematical Papyri*.

Listed are multiples of 64 from 1 times 64 up to 16 times 64, however, without any explicit reference to what they are. The text gives only a list of 16 numbers. There is no further text to motivate these entries or the whole list, nor have the multipliers been noted. The list simply ends with 1024 (i.e., 16 × 64).

It is noteworthy that this constitutes the only occurrence of a multiplication table of integers within the entire corpus of Egyptian mathematical texts. In contrast, Mesopotamia presents us with a wealth of tables for multiplications, reciprocals, and others, in which a formal tabular format is observed and sometimes even headers of individual columns are indicated.[14] Ritter convincingly explains the existence of specific types of tables in Egypt and Mesopotamia (for Middle Kingdom Egypt and Old Babylonian Mesopotamia), with their respective difficulties in distinct arithmetical areas.[15]

With fraction reckoning at the height of arithmetical techniques, it is not surprising that there are also multiplication tables of fractions. Two such tables are found in papyrus BM 10794, and it may be speculated that more of their kind might have been found on the (not extant) remainder of this source. The two extant tables, containing multiplications from 1 to 10 for the respective fractions $\frac{1}{90}$ and $\frac{1}{150}$ show the same format as indicated here for the second example ($\frac{1}{150}$). After the only partly preserved title, ten lines of multiplications and their results follow:

PAPYRUS BM 10794, NUMBER 2[16]

The way to take [. . .]

1 as $\overline{150}$

2 as $\overline{90}\ \overline{450}$

3 as $\overline{60}\ \overline{300}$

4 as $\overline{45}\ \overline{2[25]}$

5 as $\overline{30}$

6 as $\overline{30}\ [\overline{150}]$

7 as $\overline{30}\ \overline{90}\ [\overline{450}]$

8 as $\overline{20}\ \overline{300}$

9 as $\overline{30}\ [\overline{45}\ \overline{225}]$

10 as $\overline{15}$

14 For an overview of tables in Mesopotamia, see Robson, "Tables."
15 Ritter, "Measure for measure."
16 No. 67 in Parker, *Demotic Mathematical Papyri*, no. 67

There is no further evidence how the individual entries of this table were calculated. As Parker notes, several lines show additive relations to each other; that is, the entry for 6 times may result from adding the entries for times 1 and times 5, the entry for 7 times may result from adding the entries for times 2 and times 5, the entry for times 8 may result from adding the entries for times 3 and times 5 and simplifying $\overline{30}\ \overline{60}$ into $\overline{20}$, and the entry for times 9 may result from adding the entries for times 4 and times 5.[17] Similarly, the entry for times 4 easily results from doubling the entry for times 2, where both elements happen to be even—possibly the motivation for choosing the representation of $\overline{90}\ \overline{450}$ over $\overline{150}\ \overline{75}$ for $2 \times \overline{150}$. Likewise, the entries for times 3 and times 6 and those for times 5 and times 10 are linked by simple halving of the respective denominators.

The reader may have noticed that the entry for times 6 appears twice in the reasoning for the entries of the table just given—pointing to the uncertainty of assumptions like those just given: numbers can often be manipulated in more than one way to achieve a desired result. Usually, there is no way of knowing which way a scribe preferred (let alone why)—unless, of course, a significant mistake is found in one entry and then continued into another entry. As for our example, however, all entries are correct.[18]

17.2.2 Division (and a Note on Types of Fractions)

Similar to the way that divisions are carried out in the hieratic sources, the division is performed by multiplying the divisor, as can be seen in the excerpt from problem 3 of the Cairo mathematical papyrus:[19]

EXCERPT FROM PAPYRUS CAIRO, NUMBER 3

You shall take the 47th part of 100:

47 to 1

94 to 2

Remainder 6: result $\frac{6}{47}$

Result: $2\frac{6}{47}$.

Two times 47 brings the count up to 94; hence 2 is the largest integer multiplier. The difference to 100 is indicated as the remainder 6, then as $\frac{6}{47}$, and the result of the division is

---

17   Ibid., p. 73.
18   Ibid., p. 72. Note that Parker suggests similar theories for the creation of the previous table (no. 66 in his edition) for 90. However, up to times 6, the entries are not extant.
19   Ibid., no. 3.

given as $2\frac{6}{47}$. From divisions like this, where the result apparently included a nonunit fraction, it has been concluded that demotic mathematics has abandoned the former Egyptian "restriction" regarding unit fractions.[20] However, surprisingly enough, this does not hold for all demotic texts, and even within the same papyrus we find these "new fractions" and evidence of the use of the former fraction system, as evidenced by the division of $4 \div 5$ in number 9 of the Cairo papyrus, with the result given as $\overline{\overline{3}}\ \overline{10}\ \overline{30}$. David Fowler has therefore argued that the occurrences of nonunit fractions within the demotic mathematical texts are to be seen as "incomplete divisions."[21] However, if we return to number 3 and follow its course, the situation appears to be slightly more complicated. The division mentioned earlier ($100 \div 47$) was a calculation performed in the course of another division, namely, $100 \div 15\overline{\overline{3}}$. Following is the complete text of the problem:

PAPYRUS CAIRO, NUMBER. 3

If it is said to you: [Bring $15\ \overline{\overline{3}}$ to 100.]

You shall subtract 6 [from 100, remainder 94.]

You shall say [6 to 94 is 1]$5\ \overline{\overline{3}}$.

Bring it to the number $\overline{3}$, result: 47.

You shall take the number 47 to 100.

47 to 1

94 to 2

Remainder 6: result $\frac{6}{47}$

Result: $2\frac{6}{47}$.

You shall reckon its 3 times: result $6\frac{18}{47}$.

You shall say: $6\frac{18}{47}$ is its number.

You shall reckon $6\frac{18}{47}$ $15\ \overline{\overline{3}}$ times.

$6\frac{18}{47}$ to 1

$63\frac{39}{47}$ to 10

$31\ \overline{2}\ \frac{19\frac{1}{2}}{47}$ to 5

Total: $95\ \overline{2}\ \frac{11\frac{1}{2}}{47}$ to 15

$4\frac{12}{47}$ to $\overline{\overline{3}}$.

Total: $99\ \overline{2}\ \frac{23\frac{1}{2}}{47}$ to 15 <$\overline{\overline{3}}$>.

20   Ibid. For example, pp. 8–10.
21   Fowler, *Mathematics of Plato's Academy*, pp. 261–62.

Its remainder is $\overline{2}$, which is equivalent to $\frac{23\,\overline{2}}{47}$.

Total: 100.

As often witnessed in Egyptian arithmetic, the scribe begins with a clever trick, knowing that 94 ÷ 6 equals 15 $\overline{\overline{3}}$ (line 3). This, however, is no longer considered in the following solution. The scribe uses the auxiliary 47, that is, 3 times the divisor (line 4) and carries out the division with this new divisor: 100 ÷ 47, with the previously stated "result" of $2\frac{6}{47}$. To revert to the original division, this must be multiplied by 3 (as the division was carried out with 3 times the divisor), which is executed in line 10, and the result is given as $6\frac{18}{47}$, which is then (line 11) announced as the final result.

While I agree with David Fowler that $\frac{6}{47}$ is indeed an "incomplete division," it is remarkable to what degree this can be manipulated, and is accepted as part of the final answer. These "incomplete divisions" can be multiplied (as witnessed in this example when 3 times $\frac{6}{47}$ results as $\frac{18}{47}$); they are accepted as a final answer and even used in the verification. Several observations can be drawn from the series of values appearing during the verification. First of all, the divisor (or what we would call the denominator) remains as 47 throughout, even at the cost of fractions appearing in the dividend (our numerator). This is not surprising, given the history of the hieratic use of auxiliaries, which may include fractions as well.[22] If the dividend surpasses the divisor, it is automatically incorporated into the integer part of the respective number, that is, in the multiplication with 10, when $6\frac{18}{47}$ becomes $63\frac{39}{47}$. As in the hieratic texts, the multiplication by 5 is achieved as halving the result of the multiplication by 10, which had been carried out just before, and hence $63\frac{39}{47}$ becomes 31 $\overline{2}$ $\frac{19\,\overline{2}}{47}$. Note the seamless combination of fraction ($\overline{2}$) and "incomplete division" ($\frac{19\,\overline{2}}{47}$). Finally, the last incomplete division in the course of this problem ($\frac{23\,\overline{2}}{47}$) is explicitly stated to be equivalent to $\overline{2}$, and thus the verification confirms the result. In this last step, $\frac{23\,\overline{2}}{47}$ is called a remainder, thus confirming its status as a placeholder for an incomplete division.

This example may also be used to highlight how a modern-based point of view may not help our insight into an ancient mathematical culture. The evidence from the demotic mathematical papyri clearly proves that the concept of fractions did not change, which has met with varying degrees of understanding or astonishment. Consider, for example, the statement of Richard Parker:[23]

---

22   See, for example, Rhind Papyrus, no. 23, where the use of the auxiliary $\overline{45}$ results in the facilitation of the addition but still requires the scribe to work with the fractions $\overline{2}, \overline{4}, \overline{8}$ within the individual entries.

23   Parker, *Demotic Mathematical Papyri*, p. 10.

Nevertheless, despite the advantages of the new fractions, they were obviously considered as only an adjunct to unit fractions, which continued to dominate all calculations and were not abandoned even well into the Coptic period.

Compare this with the assessment of David Fowler:[24]

The contempt that is often expressed for the system of unit fraction calculation is, I believe, not entirely justified. Unit fraction expressions can be evaluated and manipulated by a wealth of algorithms, and they can convey some information, for example of magnitude and approximation, as efficiently as any other representation. The problems that do arise with unit fraction arithmetic are often exaggerated and distorted: on the one hand, unit fraction arithmetic is quite feasible, as our evidence of ancient commercial practice demonstrates abundantly, while on the other hand, the problems inherent in any formally correct and complete description of arithmetic with real numbers are far greater than many mathematicians seem to believe.

17.3 SELECTED EXAMPLES OF DEMOTIC MATHEMATICAL PROBLEMS

Like their hieratic predecessors, the demotic mathematical papyri present us with a variety of problems of various types. First examples have already been discussed in the previous section. This section presents a selection of problems from the demotic mathematical papyri focusing on those examples that occur as clusters of problems, thus enabling a more substantial analysis, because lacunae in individual problems may be filled using what we know from extant other problems of the same type.

17.3.1 Calculation of Areas

Numbers 8–12 and 14–18 of the Cairo papyrus all deal with a rectangular piece of cloth and the manipulation of its area by changing its dimensions. Given are an original height and width, one of which is then altered. The task is to calculate the other dimension under

---

24 Fowler, *Mathematics of Plato's Academy*, p. 266.

the premise that the area remains the same. This is a good example of a *suprautilitarian* problem. The way it is phrased indicates that the aim is not to hand out a specific amount of fabric to somebody, but to demonstrate the ability to manipulate rectangular areas in calculation. While there is no doubt that there may be practical applications for the mathematics involved, these do not seem to be at the center of the motivation behind this group of problems. Problem 7, just before this group of problems, is concerned with the dimensions of a sail when area and the ratio of height and width are given. This problem begins with a more elaborate introduction, part of which may serve as the heading for the following group of problems as well:

EXCERPT FROM PAPYRUS CAIRO, NUMBER 7

The things you (should) know about the articles of cloth. Viz.

If it is said to you: 'Have sailcloth made for the ships,'

and it is said to you: 'Give 1000 cloth-cubits to one sail;

have the height of the sail be (in the ratio) 1 to 1 $\overline{2}$ the width,'

(here is) the way of doing it.[25]

Again, despite the specific indication of a practical setting, I would like to argue that this also is a suprautilitarian problem based on its position within this text and also on the type of data given, that is, the area as a point of departure to calculate the dimensions of a sail. However, while I am confident that some problems are indeed practical problems originating from the scribes' work practice (see, for example the calculation of the area of rectangular fields in the last two problems of papyrus BM 10520), and there are others that are clearly suprautilitarian (as the calculation of the volume of a mast in terms of its content for an amount of water); there is also a gray area, in which a clear-cut decision is not straightforward. Problems 8–12 and 14–18 are more of a straightforward case, as removing a strip of a defined area from a piece of cloth and then determining what must be "added" to it on another side to preserve the total area would pose some practical problems. The only way to match these problems to an actual practical situation is to make the assumption that there are two pieces of cloth, the width of one being 1 cubit less in height, and one would need pieces of equal area (for the sake of equal value?). Note, however, that the problems are not phrased like this; the text mentions only one piece of cloth that is manipulated, as can be seen from the following example:

---

25   Parker, *Demotic Mathematical Papyri*, p. 19.

PAPYRUS CAIRO, NUMBER 8

The way of knowing (the) compensation of the wi[dth, when one subtracts from the height]

A measure (of cloth) which is 7 cubits (in) hei[ght and 5 cubits in width], amounting to 35 cloth-cubits.

Take off 1 (cubit) from its [height, add it to its] width.

What is that which is added to [its width]?

To cause that you know it. Viz.

The height is [7] cubits. [Subtract 1: result, its $\overline{7}$]

You shall say: '$\overline{7}$ is taken off; 6 cubits is (the) remainder; the height (is)] 6 cubits.'

Now its taken-off (area) makes [5 cloth-cubits].

You shall say: '5 is what (fraction) of 6?'

Result: its $\frac{5}{6}$. [Add it to 5: result $5\frac{5}{6}$.]

The width (is) $5\frac{5}{6}$ (cubits).

You shall reckon [$5\frac{5}{6}$, 6 times: result: 35 cloth-cubits],

which will make the (given)-number.[26]

Richard Parker interpreted the solution as determining "what fraction of the reduced height the cut-off strip would be, and then adds this to the width."[27] While the procedures of problems 8, 9, and 10 indeed determine the cutoff strip as a fraction of the original height, the algorithms reveal that this does not play a further part in the solution of the problem. The algorithm of the procedure (in its numerical and its abstract version) can be written as follows:

	7		D_1
	5		D_2
	35		$D_1 \times D_2$
	1		D_3
(1)	$7 - 1 = 6$	(1)	$D_1 - D_3$
(2)	$1 \div 7 = \overline{7}$	(2)	$D_3 \div D_1$
(3)	$1 \times 5 = 5$	(3)	$D_3 \times D_2$
(4)	$5 \div 6 = \frac{5}{6}$	(4)	$(3) \div (1)$
(5)	$5 + \frac{5}{6} = 5\frac{5}{6}$	(5)	$D_2 + (4)$
(v)	$5\frac{5}{6} \times 6 = 35$	(v)	$(5) \times (1) = D_1 \times D_2$

---

26 Ibid., p. 20.
27 Ibid., p. 20.

The method of rewriting the rhetorical text in the form of an algorithm helps to follow the structure of the procedure. The procedure determines the new height (1), as well as the fraction of the cutoff part as indicated by Parker (2), but then moves on to calculate the area of the cutoff piece as the product of cutoff length and original width (3). To determine the additional width, the area of the cutoff piece is divided by the new height (4). (5) adds the original width to this to obtain the new width. The fractional part of the reduced height, which was determined in (2), does not appear again within the algorithm.

The following problem 9 is a variant of problem 8, the piece of cloth has a height of 6 cubits and a width of 4 cubits; again, the area is indicated explicitly as 24 cloth-cubits. One cubit is taken off the height, and the new width is calculated. The translation of the procedure reads as follows:

EXCERPT FROM PAPYRUS CAIRO, NUMBER 9[28]

The height (is) 6. Viz.

Subtract 1: result, its $\overline{6}$.

You shall say: "[$\overline{6}$] is taken off; 5 cubits is (the) remainder;

(the) height (is) 5 cubits."

Now [its] taken-off (area) makes 4 cloth-cubits.

You shall say: "4 is what (fraction) of 5?"

[Result: its $\overline{\overline{3}}$] $\overline{10}\ \overline{30}$.

Add it to 4: result 4 $\overline{\overline{3}}\ \overline{10}\ \overline{30}$. The number is the width.

To cause that you know it. Viz.

You shall reckon 4 $\overline{\overline{3}}\ \overline{10}\ \overline{30}$, 5 times: result: 24 cloth-cubits,

which will make (the given)-number again.

From this procedure, the following algorithm can be established:

6	D_1
4	D_2
24	$D_1 \times D_2$
1	D_3
(1) $6 - 1 = 5$	(1) $D_1 - D_3$
(2) $1 \div 6 = \overline{6}$	(2) $D_3 \div D_1$

---

28   Ibid., p. 21.

(3)	$1 \times 4 = 4$		(3)	$D_3 \times D_2$
(4)	$4 \div 5 = \overline{\overline{3}} \, \overline{10} \, \overline{30}$		(4)	$(3) \div (1)$
(5)	$4 + \overline{\overline{3}} \, \overline{10} \, \overline{30} = 4 \, \overline{\overline{3}} \, \overline{10} \, \overline{30}$		(5)	$D_2 + (4)$
(v)	$4 \, \overline{\overline{3}} \, \overline{10} \, \overline{30} \times 5 = 24$		(v)	$(5) \times (1) = D_1 \times D_2$

The comparison of the most abstract form of the algorithm reveals the identity of the procedures of problems 8 and 9.

The following problem (10) shows a modification of this basic algorithm to facilitate the handling of fractions. The text of the problem reads as follows:

Papyrus Cairo, Number 10

A band which is 6 cloth-cubits (in) [hei]ght and (in) width [$1\overline{2}$ cubits, amounting to 9 cloth-cubits]. Take it, $\overline{2}$ cubit, off its height and add [it to its width]. What is its equivalent?

Now (the) hei[ght is 6]. Take it, $\overline{2}$ from it: <result>: its $\overline{12}$.

You shall say: "$\overline{12}$ is taken off; 11 cubits [is (the) remainder; (the) height (is) 11 cubits]."

Now the taken-off (area) makes $1\overline{2}$ cloth-cubits.

[You shall say: "3 is what (fraction) of 11?"]

Result: $\frac{3}{11}$.

Add it to 3: result $3\frac{3}{11}$ [You shall reckon $3\frac{3}{11}$], $\overline{2}$ times: result $1\,\overline{2}\,\frac{12}{11}$.

[It is the width.]

[To cause that you know it. Viz.]

[You shall reckon $1\,\overline{2}\,\frac{12}{11}$, $5\,\overline{2}$ times: result: 9 cloth-cubits]

[which will make (the given)-number again.][29]

At first glance, that is, when established as in the previous problems, the algorithm looks as follows:

6		D_1
$1\overline{2}$		D_2
9		$D_1 \times D_2$
$\overline{2}$		D_3

---

29   Ibid., p. 22.

(1)	$6 - \overline{2} = 5\,\overline{2}$	(1)	$D_1 - D_3$
(2)	$\overline{2} \div 6 = \overline{12}$	(2)	$D_3 \div D_1$
(3)	$5\,\overline{2} \times 2 = 11$	(3)	$(1) \times 2$
(4)	$[\overline{2} \times 2 = 1]$	(4)	$[D_3 \times 2]$
(5)	$1 \times 1\overline{2} = 1\,\overline{2}$	(5)	$(4) \times D_2$
(6)	$1\,\overline{2} \times 2 = 3$	(6)	$D_2 \times 2$
(7)	$3 \div 11 = \frac{3}{11}$	(7)	$(6) \div (5)$
(8)	$3 + \frac{3}{11} = 3\frac{3}{11}$	(8)	$(6) + (7)$
(9)	$3\frac{3}{11} \times \overline{2} = 1\,\overline{2}\frac{12}{11}$	(9)	$(8) \times \overline{2}$
(v)	$1\,\overline{2}\frac{12}{11} \times 5\,\overline{2} = 9$	(v)	$(9) \times (1) = D_1 \times D_2$

Although the number of steps carried out in this procedure is almost double that of the previous problems, the procedures are, in fact, very similar, as can be revealed by renumbering the additional steps with an asterisk (*), as follows:

	6		D_1
	$1\,\overline{2}$		D_2
	9		$D_1 \times D_2$
	$\overline{2}$		D_3
(1)	$6 - \overline{2} = 5\,\overline{2}$	(1)	$D_1 - D_3$
(2)	$\overline{2} \div 6 = \overline{12}$	(2)	$D_3 \div D_1$
(*1)	$5\,\overline{2} \times 2 = 11$	(*1)	$(1) \times 2$
(*2)	$[\overline{2} \times 2 = 1]$	(*2)	$[D_3 \times 2]$
(3)	$1 \times 1\overline{2} = 1\,\overline{2}$	(3)	$(*2) \times D_2$
(*3)	$1\,\overline{2} \times 2 = 3$	(*3)	$D_2 \times 2$
(4)	$3 \div 11 = \frac{3}{11}$	(4)	$(*3) \div (*1)$
(5)	$3 + \frac{3}{11} = 3\frac{3}{11}$	(5)	$(*3) + (4)$
(*5)	$3\frac{3}{11} \times \overline{2} = 1\overline{2}\frac{12}{11}$	(*5)	$(5) \times \overline{2}$
(v)	$1\overline{2}\frac{12}{11} \times 5\,\overline{2} = 9$	(v)	$(*5) \times (1) = D_1 \times D_2$

17.3.2 The Pole-against-the-Wall Problems

For the later Egyptian mathematical texts, it has been claimed that there is a transmission of knowledge from Mesopotamia or Greece since their first publication. The example par excellence for this claim has always been the pole-against-the-wall problem—a problem known from two Mesopotamian sources, one Old Babylonian (number 9 of tablet BM 85196), the other Seleucid (number 12 of tablet BM 34568), and from one demotic source (the Cairo demotic mathematical papyrus), which has a series of pole-against-the-wall problems. However, this is the only very distinct case, in which a problem existed in Old Babylonian times, still exists in Seleucid sources, and makes its first appearance in the Egyptian material of the Greco-Roman Period.[30] The lack of evidence between the two periods, however, for Egypt as well as for Mesopotamia, makes it impossible to definitely prove a specific transmission, let alone when and how it happened.

The oldest attestation of the problem originates from Mesopotamia in form of problem 9 of the tablet BM 85196.[31] The first difference between any Egyptian and Mesopotamian mathematical text is the number system. Whereas Egypt used a nonpositional decimal system, Mesopotamian mathematics used sexagesimal positional number notation. These are rendered in the translation by Høyrup through indicating the first position after the sexagesimal comma by ' and the second postition by "; thus 30' in this sexagesimal notation is 0.5 in decimal notation.[32] Throughout the Mesopotamian examples, the sexagesimal notation has been kept; however, in those instances where constants appear in the abstract algorithms, they have been rendered decimally in order to facilitate a comparison with the demotic algorithms.

The translation of this problem given by Høyrup reads as follows:

BM 85196, Number 9

A pole, 30', (that is,) a reed, from [...] its [...]

Above, 6' it has descended, be[l]o[w, what has it moved away?]

You, 30' make hold, 15' you see.

6' fro[m] 30' te[ar out, 24' you see.]

24 make hold, 9' 36" you see,

9' 36" from [15' tear out], 5' 24" you see.

5' 24", what is equalside?

[18' is equalside, 18'] on the ground it has moved away.

If 18' o[n the g]round, above, what did it descend?

---

30 For a recent discussion of the individual attestations of this problem, see Melville 2004.
31 Originally published in Neugebauer, *Mathematische Keilschrift-Texte II*, pp. 43–46. For a recent translation of problems from this tablet, see Høyrup, *Lengths, Widths, Surfaces*. For the later spread of this problem, see Sesiano, "Survivance."
32 This is a slightly simplified version of that used by Høyrup, indicating only the fractional positions with special marks and the positive powers of 60 by separating them with commas. On various conventions for rendering sexagesimal numbers, see Høyrup, *Lengths, Widths, Surfaces*, p. 12.

18′ make hold, 5′ 24″ you see,

5′ 24″ from 15′ tear out, 9′ 36″ you see.

9′ 36″, what is equalside?

24′ is equalside,

24′ from 30′ tear out, 6′ you see (for what) it has descended.

Thus the procedure.[33]

A pole of length l, which is placed vertically against a wall, is then moved into a slanted position, in which its top descends by the length d, and its foot moves a distance s away from the wall (see figure 33). The first part of the procedure uses l and d to calculate s. The second part of the problem then presents the reverse calculation, using l and s to calculate d (a verification for the correctness of the result for s).

The symbolic algorithm for this procedure looks as follows.

First part of the procedure

	30′	D_1
	6′	D_2
(1)	$30'^2 = 15'$	(1) D_1^2
(2)	$30' - 6' = 24'$	(2) $D_1 - D_2$
(3)	$24'^2 = 9'\ 36''$	(3) $(2)^2$
(4)	$15' - 9'\ 36'' = 5'\ 24''$	(4) $(1) - (3)$
(5)	$\sqrt{5'\ 24''} = 18'$	(5) $\sqrt{(4)}$

Second part of the procedure (using D_1 and (5) as data)

(6)	$(18')^2 = 5'\ 24''$	(6) $(5)^2$
(7)	$15' - 5'\ 24'' = 9'\ 36''$	(7) $(1) - (6)$
(8)	$\sqrt{9'\ 36''} = 24'$	(8) $\sqrt{(7)}$
(9)	$30' - 24' = 6'$	(9) $D_1 - (8)$

Both parts are a "straightforward application of the Pythagorean rule."[34] I use the distinction between the Pythagorean theorem (i.e., Euclid I.47) and the Pythagorean rule as

---

33   Ibid., p. 275.
34   Ibid., p. 275.

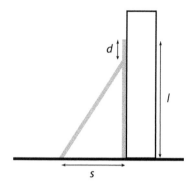

FIGURE 33: The pole-against-the-wall problem (drawn after Høyrup, *Lengths, Widths, Surfaces*, p. 276, figure 79)

established by Jens Høyrup.[35] The Seleucid problem is problem 12 in a tablet (BM 34568) of about twenty problems, most of which are concerned with rectangles.[36] Høyrup translates the problem as follows:

BM 34568, NUMBER 12

One reed together with a wall I have erected.

3 *kuš* as much as I have gone down, 9 *kuš* it has moved away.

What the reed? What the wall?

Since you do not know,

3 steps of 3: 9. 9 steps of 9: 1,21.

9 to 1,21 y[ou jo]in[:]

1,3[0 steps of 3]0 you go: 45.

Igi 3: 20. 20 [steps of 45 you go:]15 the reed.

Wh[at the wa]ll?

15 steps of 15: 3,45. 9 [steps of 9]: 1,21.

1,21 from 3,45 you lift, remaining [2,24].

What steps of what may I go so that 2,[24]?

12 steps of 12: 2,24. 12 the wall.[37]

35 Høyrup, "Pythagorean rule and theorem."
36 Melville, "Poles and walls," p. 153. The tablet was originally published in Neugebauer, *Mathematische Keilschrift-Texte III*, pp. 14–17. For a recent translation, see Høyrup, *Lengths, Widths, Surfaces*, pp. 391–99.
37 Høyrup, *Lengths, Widths, Surfaces*, p. 394.

The situation dealt with in this problem is that of a reed, which is at first placed vertically against a wall. Then it is moved so that the top has descended 3 cubits, the base has moved 9 cubits from the wall, and the top of the reed is now resting on top of the wall. The problem asks us to determine the length of the reed and the height of the wall. Rewritten in form of the symbolic algorithm, the procedure looks as follows:[38]

First part of the procedure: length of the reed

	3		D_1
	9		D_2
(1)	$3 \times 3 = 9$	(1)	$D_1 \times D_1$
(2)	$9 \times 9 = 1, 21$	(2)	$D_2 \times D_2$
(3)	$9 + 1, 21 = 1, 30$	(3)	$(1) + (2)$
(4)	$1, 30 \times 30' = 45$	(4)	$(3) \times \frac{1}{2}$
(5)	$1 \div 3 = 20'$	(5)	$1 \div D_1$
(6)	$20' \times 45 = 15$	(6)	$(5) \times (4)$

Second part of the procedure: height of wall

(7)	$15 \times 15 = 3, 45$	(7)	$(6) \times (6)$
(8)	$9 \times 9 = 1, 21$	(8)	$D_2 \times D_2$
(9)	$3, 45 - 1, 21 = 2, 24$	(9)	$(7) - (8)$
(10)	$\sqrt{2, 24} = 12$	(10)	$\sqrt{(9)}$

Let us compare these two Mesopotamian problems with some examples from the demotic pole-against-the-wall problems (papyrus Cairo, numbers 24–31). The problems present varying scenarios: In numbers 24–26, the foot of a pole of given height is moved outward a certain distance, resulting in the lowering of the top, and the amount of this lowering is then determined. Numbers 27–29 determine how far the foot of a pole of given length is moved outward, when the top is lowered a certain amount; and numbers 30–31 determine the length of the pole for a given distance that its foot is moved outward and the amount the top is lowered in the process.

PAPYRUS CAIRO, NUMBER 27:

A pole that is 10 cubits [when erect].

The number of cubits to bring [out its foot] will not be stated.

Lower its top [2] cubits.

Wh[at is the num]ber of [its foot (moved) outward from it?]

You shall reckon 10, 10 times: result [1]00.

Take it, [2, from 10]: remainder 8.

You shall reckon 8, 8 times: result 64.

Take it from 100: remainder [36].

Cause that [it] reduce to its square root: [result 6].

You shall say:

'6 is [the] number of its foot (moved) outward from it.'[39]

The procedure can be rewritten as follows:

	10		D_1
	2		D_2
(1)	$10 \times 10 = 100$	(1)	$D_1 \times D_1$
(2)	$10 - 2 = 8$	(2)	$D_1 - D_2$
(3)	$8 \times 8 = 64$	(3)	$(2) \times (2)$
(4)	$100 - 64 = 36$	(4)	$(1) - (3)$
(5)	$\sqrt{36} = 6$	(5)	$\sqrt{(4)}$

The comparison of the abstract algorithm of number 27 of papyrus Cairo with that of number 9 of the Old Babylonian tablet BM 85196 reveals the identity of their procedures:

BM 85196, Number 9		Cairo Papyrus, Number 27	
	D_1		D_1
	D_2		D_2
(1)	D_1^2	(1)	$D_1 \times D_1$
(2)	$D_1 - D_2$	(2)	$D_1 - D_2$
(3)	$(2)^2$	(3)	$(2) \times (2)$
(4)	$D_1 - (3)$	(4)	$D_1 - (3)$
(5)	$\sqrt{(4)}$	(5)	$\sqrt{(4)}$

---

39   Translation from Parker, *Demotic Mathematical Paypri*, pp. 37–38.

The formal difference in the operations of the first and third steps is caused by the respective underlying arithmetic systems. Demotic mathematics phrases the operation of squaring as a multiplication of two identical numbers.

While problem 27 is a match to the Old Babylonian version of the pole-against-the-wall problem, problem 31 provides a match to the Seleucid version:

PAPYRUS CAIRO, NUMBER 31

[A pole, if the] number of its foot (moved) outward is 10 cubits [and the number of
  lowering its top is] 4 [cubits], what was its peak?
[You shall reckon 10, 10 times: result 1]00.
You shall reckon 4, 4 times: result: 16.
[You shall add it to] 16: result 116.
You shall add 4 to 4: result 8.
[Take it to] 116: result 14 $\overline{2}$.
The peak of the pole was 14 [$\overline{2}$].[40]

Again, rewritten in the form of its symbolic algorithm, the procedure looks as follows:

	10		D_1
	4		D_2
(1)	$10 \times 10 = 100$	(1)	$D_1 \times D_1$
(2)	$4 \times 4 = 16$	(2)	$D_2 \times D_2$
(3)	$100 + 16 = 116$	(3)	$(1) + (2)$
(4)	$4 + 4 = 8$	(4)	$D_2 + D_2$
(5)	$116 \div 8 = 14\,\overline{2}$	(5)	$(3) \div (4)$

A comparison of the abstract algorithms shows that the situations described in the problems are identical; however, the procedures that are used for their solution differ slightly in two respects on the arithmetical level. Note, however, that in this example, the Mesopotamian procedure also uses a multiplication of two identical numbers to express squaring.

---

40   Ibid., pp. 39–40.

Seleucid Tablet, Number 12	Papyrus Cairo, Number 31
D_1	D_1
D_2	D_2
(1) $D_1 \times D_1$	(1) $D_1 \times D_1$
(2) $D_2 \times D_2$	(2) $D_2 \times D_2$
(3) (1) + (2)	(3) (1) + (2)
(4) (3) $\times \frac{1}{2}$	(4) $D_2 + D_2$
(5) $1 \div D_1$	(5) (3) ÷ (4)
(6) (5) × (4)	

The first difference between the algorithms has already been noted in the previous example: Mesopotamian mathematics does not use divisions as arithmetic operations. Instead, the inverse of the divisor is calculated and then multiplied by the dividend. Therefore, the Mesopotamian procedure of the problem consists of six steps, whereas the demotic procedure only has five.

The second difference appears in the fourth step. The Mesopotamian version shows the operation of halving (or multiplication by $\frac{1}{2}$), whereas the demotic procedure doubles the second datum, D_2 (shown as an addition: $D_2 + D_2$).

This difference can be grasped easily by a modern reader when comparing the procedures with the algebraic solution that a modern reader is used to, which will be briefly sketched. If L designates the length of the pole (the unknown of this problem), D_2 is the amount that the pole was lowered, and D_1 is the amount that the foot of the pole moved away from the wall, using the Pythagorean theorem, we might start with the equation

$$(L - D_2)^2 + D_1^2 = L^2.$$

Solving this equation for L yields

$$L = \frac{D_1^2 + D_2^2}{2D_2}.$$

The difference between the procedures can be understood looking at the denominator of our modern solution. The 2 of the denominator can be found in the Mesopotamian procedure as a factor $\frac{1}{2}$, by which the sum of the squares of D_1 and D_2 (found as the numerator of our equation) are multiplied. The demotic procedure implements the doubling of D_2 as an addition ($D_2 + D_2$).

Therefore, while the general strategy of the two procedures in this case may be related and can be described as using the Pythagorean rule, the individual procedures are distinct from each other on the arithmetic level.

18.

Conclusion:
Egyptian Mathematics
in Historical Perspective

The evidence presented in the previous chapters leaves little doubt that mathematical practices played a central role in Egyptian culture at all times. And, like today, much of it surfaced only indirectly through a display of numeracy of some kind. It has been suggested that in the beginning, it was numeracy that drove literacy, and at least two cultures with extant evidence for their earliest writing (i.e., Egypt and Mesopotamia) provide evidence for this claim.[1] Even the earliest extant written evidence from Egypt includes numbers. In Egypt, the use of writing and numbers was a prerogative only of the elite. This usage included, on the one hand, the management of wealth, which then led to the creation and development of metrological systems, and later on (with the earliest evidence dating from the Middle Kingdom) to a corpus of mathematical techniques to be employed by the administrative body ("the scribes"), which allowed control of available resources with utmost efficiency (as exemplified in the administrative problems in the hieratic mathematical papyri). On the other hand, however, numbers have had a symbolic use, which also can be documented almost from the very beginning. They were used to represent power and have a place within magic and religious contexts as well.[2] While not all numbers extant in ancient sources involve mathematics, it is remarkable how often at least some basic technique used to organize and manipulate these numbers is involved. A comparison between the mathematical techniques of Egypt and Mesopotamia can

---

1   See Benoit, Chemla, and Ritter, *Histoire de fractions*, pp. 3–4, and Ritter, "Measure for measure," p. 45.
2   For a case study of the magic use of the number 7, see Rochholz, *Symbolgehalt*.

be used to highlight differences, which prove that mathematics is a cultural activity depending on the culture that developed and used it. Jens Høyrup has pointed out a significant difference between Egypt and Mesopotamia in the motivation of using mathematics:[3]

> The first record of Pharaonic-Egyptian mathematics is the Early Dynastic or Late Predynastic Macehead of King Narmer (ca. 3500 BCE), on which a tribute to the king of 400,000 bulls, 1,422,000 goats and 120,000 prisoners is represented numerically. This illustrates well that the legitimacy of the Pharaonic state was derived from conquest (and the promise of cosmic stability), unlike that of the early Mesopotamian state, which was based on mathematical justice.

Nevertheless, mathematics became a state-directed activity in both cultures, which was put to use in the administration of the country. Due to different number systems and, consequently, different arithmetic, each culture developed its own mathematics, as was pointed out in a comparison of two contemporary Egyptian and Mesopotamian mathematical problems of the same topic by Jim Ritter:[4]

> By the early second millennium, the two cultures each had a sophisticated and efficient mathematics in place, which could be and was applied to the problems of society. Since both cultures at this time were highly centralized bureaucratic states, with an intensive, irrigation based agriculture and a developed inland and foreign trade, one might expect that the problems treated would be similar. Indeed, the mathematics of both cultures are generally treated as 'practical' or 'empirical' by almost all historians of mathematics, and to a certain extent this is true. . . . But a closer look at the mathematics during, say, the first half of the second millennium in the two civilizations will show that there were in place two mathematics, so that even where the 'same' problem is treated, the methods used are quite different.

Mesopotamian arithmetic, based on a sexagesimal number system, used multiplication with an inverse instead of a direct division. Its main technical problem arose from the possibility of arriving at a nonfinite sexagesimal number when calculating an inverse or a square root.

---

3   Høyrup, "Mathematics," p. 465.
4   Ritter, "Measure for measure," pp. 48–49.

Therefore, tables of inverses and roots were an absolute necessity.[5] In addition, multiplication tables were used to facilitate the carrying out of multiplications with integers. Egypt, however, developed a method for carrying out multiplications and divisions in writing. Thus, it is not surprising that no multiplication tables are extant from the earlier hieratic corpus of mathematical texts. Instead, the mathematical texts include tables for the handling of fractions (most notably the $2 \div n$ table) and metrological conversions. The later demotic mathematical corpus shows several multiplication tables, including one for an integer (64) and two for fractions ($\overline{90}$ and $\overline{150}$), possibly reflecting an exchange of ideas with Mesopotamia and the fact that handling fractions remained technically demanding.

Another major difference between the two cultures, which affects our perception of them, originated from the writing material used. While papyrus was developed in Egypt early on to serve as the writing material for daily life documents, the scribes of Mesopotamia used clay tablets, in which they pressed cuneiform signs with a reed stylus. As a consequence, vast numbers of cuneiform texts on clay tablets are still extant today, whereas most of the Egyptian papyri have perished.[6] Despite only twenty-five mathematical texts being extant from all of Egyptian history, it is still possible to get an idea of Egyptian mathematics and its role within Egyptian culture. This necessitates the use of additional sources, some of which have been introduced in this book. Evidence for the existence of metrological systems is abundant from the Old Kingdom on. Its presence throughout Egyptian history and the changes the metrological systems underwent prove the ongoing use of mathematics in administration, but can also serve as indicators that mathematics itself changed (and thereby refute the claim that the mathematical development stagnated).

However, mathematics also possessed a place within the intellectual history of ancient Egypt, as the literary texts from the New Kingdom reveal. If we are to believe papyrus Anastasi I, a scribe proved himself a "true scribe" not only by his expertise in writing in a formally correct manner, but also by his knowledge in other subjects, such as geography and mathematics. The continued use of numbers in cultic and magic contexts further documents this aspect of mathematics. While the continuous importance of mathematics may therefore be assumed, its actual content changed, as can be seen from a comparison of the hieratic and demotic mathematical papyri. Despite similarities of the two corpora, such as the formal style of presenting mathematics rhetorically and in the form of procedures with concrete numerical

---

5   Ibid., p. 68.
6   Of the early period of state formation in Mesopotamia alone, approximately 120,000 clay tablets are extant; see Englund, *Decipherment*, p. 2.

values, there are also some differences in techniques, such as the new way of handling fractions and new techniques for carrying out multiplications. Further differences can be seen in the topics of the individual problems: While the hieratic papyri have abundant examples of practical problems, such as the calculation of the volume of a granary, the exchange of bread and beer, and the calculation of work and rations, the demotic papyri are full of so-called suprautilitarian problems, often phrased as practical problems but without a direct practical application. This change in content goes along with a change in its location; the hieratic mathematical texts can be shown in at least one case (the Lahun papyri) to be linked to Egyptian settlements. It seems reasonable to assume that scribes learned their mathematics to serve the administration of the king.

The demotic mathematical texts may be linked with temples, the last stand of Egyptian native culture.[7] Written by the scribes employed in these temples, they not only contain those problems that were still relevant, that is, the calculation of the areas of fields, but also problems that must have served, as in the case of papyrus Anastasi I, as signature knowledge for a true scribe.

Evidence presented in this book was selected to show individual aspects of Egyptian mathematics over several thousand years. A different selection may highlight further aspects of the same picture. May many more of these be painted in the future.

---

7   For a similar assessment of demotic texts in general, see Thissen, "Demotische Literatur," p. 92.

Bibliography

Abdulaziz, Abdulrahman A. "On the Egyptian method of decomposing 2/n into unit fractions," *Historia Mathematica* 35 (2008): 1–18. (Abdulaziz, "Egyptian method")

Adams, Barbara, and Cialowicz, Krzysztof. *Protodynastic Egypt*. Buckinghamshire: Shire Publications 2008. (Adams/Cialowicz, *Protodynastic Egypt*)

Allen, James P.. *The Art of Medicine in Ancient Egypt*. New York: Metropolitan Museum of Art and New Haven/London: Yale University Press, 2005. (Allen, *Medicine*)

——. *The Heqanakht Papyri* (Publications of the Metropolitan Museum of Art Egyptian Expedition 27). New Haven and London: Yale University Press, 2004. (Allen, *Heqanakht*)

——. "The high officials of the early Middle Kingdom." In Nigel Strudwick and John Taylor, eds., *The Theban Necropolis: Past, Present, and Future*. London: British Museum, 2003: 14–29. (Allen, "High officials")

——. *Middle Egyptian. An Introduction to the Language and Culture of Hieroglyphs*. Cambridge: Cambridge University Press, 2000. (Allen, *Middle Egyptian*)

——. "Review of Marshall Clagett: Ancient Egyptian science. A source book." Vol. 3: Ancient Egyptian Mathematics (Memoirs of the American Philosophical Society 232). Philadelphia: American Philosophical Society 1999," *Isis* 92 (2001): 151–52. (Allen, "Review Clagett")

Allen, Thomas: *The Book of the Dead or Going Forth by Day. Ideas of the Ancient Egyptians Concerning the Hereafter as Expressed in Their Own Terms* (Studies in Ancient Oriental Civilizations 37). Chicago: The University of Chicago Press, 1974. (Allen, *Book of the Dead*)

Andrews, Carol A. R. "Unpublished demotic papyri in the British Museum." In: *Acta Demotica. Acts of the 5th International Conference for Demotists*. Pisa: Giardini, 1994: 29–37 (Andrews, "Unpublished demotic papyri")

Arnold, Dieter. *Building in Egypt: Pharaonic Stone Masonry*. New York and Oxford: Oxford University Press, 1991. (Arnold, *Building*)

Baer, Klaus. "A note on Egyptian units of area in the Old Kingdom." *Journal of Near Eastern Studies* 15 (1957): 113–17 (Baer, "Egyptian units of area")

Baer, Klaus. *Rank and Title in the Old Kingdom: The Structure of the Egyptian Administration in the Fifth and Six Dynasties*. Chicago: University of Chicago Press, 1960. (Baer, *Rank and title*)

Baines, John. "Birth of writing and kingship: Introduction." In Béatrix Midant-Reynes and Yann Tristant, eds., *Egypt at Its Origins 2. Proceedings of the International Conference "Origin of the State, Predynastic and Early Dynastic Egypt"; Toulouse (France), September 5–8, 2005* (Orientalia Lovaniensia analecta 172). Leuven: Peeters, 2008, 841–49 (Baines, "Writing and kingship")

——. "Communication and display: The integration of early Egyptian art and writing." *Antiquity* 63 (1989): 471–82. (Baines, "Communication and display")

——. "The earliest Egyptian writing: development, context, purpose." In Stephen D. Houston ed., *The First Writing. Script Invention as History and Process*. Cambridge: Cambridge University Press, 2004, 150–89 (Baines, "Earliest Egyptian writing")

——. "Literacy, social organization, and the archaeological record: The case of early Egypt." In J. Gledhill, B. Bender, M. T. Larsen, eds., *State and Society. The Emergence and Development of Social Hierarchy and Political Centralization*. London and New York: Routledge, 1988, 192–214 (Baines, "Literacy")

———. "Writing: Invention and early development." In Kathryn A. Bard, ed., *Encyclopedia of the Archaeology of Ancient Egypt*. London and New York: Routledge, 1999, 882–85 (Baines, "Writing: Invention and early development")

Bard, Kathryn A. "The Egyptian predynastic: A review of the evidence." *Journal of Field Archaeology* 21 (1994): 265–88. (Bard, "Egyptian predynastic")

———. "The emergence of the Egyptian state (ca. 3200–2686 BC)." In Ian Shaw, ed., *The Oxford History of Ancient Egypt*. Oxford: Oxford University Press, 2000, 61–88 (Bard, "Emergence of the Egyptian state")

———. *From Farmers to Pharaohs: Mortuary Evidence for theRise of Complex Society in Egypt* (Monographs in Mediterranean Archaeology 2). Sheffield: Sheffield Academic Press, 1994. (Bard, *Farmers to pharaohs*)

———. *An Introduction to the Archaeology of Ancient Egypt*. Oxford: Blackwell, 2008. (Bard, *Archaeology of ancient Egypt*)

———. "Origins of Egyptian writing," In: Renee Friedman and Barbara Adams (ed.), *The Followers of Horus. Studies dedicated to Michael Allen Hoffman*. Oxford: Oxbow Books 1992: 297–306 (Bard, "Origins of Egyptian writing")

Baud, Michel. "Review of Toby A.H. Wilkinson, Royal Annals of Ancient Egypt," *Chronique d'Egypte* 78 (2003): 145–148 (Baud, "Review Wilkinson")

Benoit, Paul, Chemla, Karine, and Ritter, Jim. *Histoire de fractions, fractions d'histoire*. Basel, Boston, Berlin: Birkhäuser, 1992. (Benoit/Chemla/Ritter, *Histoire de fractions*)

Berger, Suzanne. "A note on some scenes of land measurement." *Journal of Egyptian Archaeology* 20 (1934): 54–56 (Berger, "Land measurement")

Berggren, John Lennart. *Episodes in the Mathematics of Medieval Islam*. New York: Springer, 1986. (Berggren, *Mathematics of Medieval Islam*)

Betlyon, John W. "Egypt and Phoenicia in the Persian Period: Partners in trade and rebellion." In Gary N. Knoppers and Antoine Hirsch, *Egypt, Israel and the Ancient Mediterranean World. Studies in honour of Donald B. Redford* (Probleme der Ägyptologie 20). Leiden: Brill, 2004, 455–77 (Betlyon, "Egypt and Phoenicia in the Persian Period")

Berlev, Oleg D. "Review W. K. Simpson, papyrus Reisner III." *Bibliotheca Orientalis* 28 (1971): 324–26. (Berlev, "Review Simpson")

Bierbrier, Morris L. *The Tomb-Builders of the Pharaohs*. London: British Museum Publications, 1982. (Bierbrier, *Tomb-builders*)

Blackman, Aylward M. *The Rock Tombs of Meir IV* (Archaeological Survey of Egypt 25). London: Egypt Exploration Society, 1924. (Blackman, *Meir IV*)

Booth, Charlotte. *The Hyksos Period in Egypt*. Princes Risborough: Shire, 2005. (Booth, *Hyksos Period*)

Borchardt, Ludwig. "Besoldungsverhältnisse von Priestern im mittleren Reich." *Zeitschrift für ägyptische Geschichte und Altertumskunde* 40 (1902/1903): 113–17. (Borchardt, "Besoldungsverhältnisse")

———. "Der zweite Papyrusfund von Kahun und die zeitliche Festlegung des mittleren Reiches der ägyptischen Geschichte." *Zeitschrift für ägyptische Sprache und Altertumskunde* 37 (1899): 89–103. (Borchardt, "Papyrusfund")

Bowman, Alan K. *Egypt After the Pharaohs (332 BC–AD 642)*. London: British Museum Publications, 2000. (Bowman, *Egypt after the Pharaohs*)

Breasted, James H. *Ancient Records of Egypt: Historical Documents from the Earliest Times to the Persian Conquest*, collected, edited and translated with commentary by James Henry Breasted. Chicago: University of Chicago Press, 1906–1907. (Breasted, *Records of Egypt*)

———. *The Edwin Smith Surgical Papyrus* (The University of Chicago Oriental Institute Publication 3). Chicago: The University of Chicago Press, 1930. (Breasted, *Edwin Smith Surgical Papyrus*)

Breyer, Francis Amadeus Karl. "Die Schriftzeugnisse des prädynastischen Königsgrabes Uj in Umm el-Qaab. Versuch einer Neuinterpretation." *Journal of Egyptian Archaeology* 88 (2002): 53–65. (Breyer, "Schriftzeugnisse")

Bruckheimer, M., and Salomon, Y. "Some Comments on R. J. Gillings' Analysis of the 2/n Table in the Rhind Papyrus." *Historia Mathematica* 4 (1977): 445–52. (Bruckheimer/Salomon, "Some Comments")

Brugsch, Heinrich. "Ein altägyptisches Rechenexempel." *Zeitschrift für Ägyptische Sprache und Altertumskunde* 3 (1865): 65–70 (Brugsch, "Rechenexempel")

Bruins, Evert M. "The part in ancient Egyptian mathematics." *Centaurus* 19 (1975): 241–51. (Bruins, "The part")

———. "Reducible and trivial decompositions concerning Egyptian arithmetics," *Janus* 69 (1981): 281–97. (Bruins, "Reducible and trivial decompositions")

Brunner, Hellmut. *Altägyptische Erziehung*. Wiesbaden: Harrassowitz, 1957. (Brunner, *Erziehung*)
——. "Schreibunterricht und Schule als Fundament der ägyptischen Hochkultur," In. Lenz Kriss-Rettenbeck and Max Liedtke, eds., *Schulgeschichte im Zusammenhang der Kulturentwicklung*. Bad Heilbrunn: Klinkhardt, 1983, 62–75. (Brunner, "Schreibunterricht und Schule")
——. *Die Texte aus den Gräbern der Herakleopolitenzeit von Siut* (Ägyptologische Forschungen 5). Glückstadt, Hamburg, New York: J. J. Augustin, 1937. (Brunner, *Herakleopolitenzeit*)
Burkhard, Günter, and Fischer-Elfert, Hans-Werner. *Ägyptische Handschriften*, Teil 4, herausgegeben von Erich Lüddeckens (Verzeichnis der orientalischen Handschriften in Deutschland, Band XIX,4). Stuttgart: Franz Steiner Verlag, 1994. (Burkhard/Fischer-Elfert, *Ägyptische Handschriften*)
Callender, Gae. "The Middle Kingdom renaissance (ca. 2055–1650 BC)." In Ian Shaw, ed., *The Oxford History of Ancient Egypt*. Oxford: Oxford University Press 2000, 148–83. (Callender, "Middle Kingdom")
Caminos, Ricardo A. *Late-Egyptian Miscellanies* (Brown Egyptological Studies I) London: Oxford University Press, 1954. (Caminos, *Late Egyptian Miscellanies*)
Caveing, Maurice. "The arithmetic status of the Egyptian quantieme." In Paul Benoit, Karine Chemla, and Jim Ritter, eds., *Histoire des fractions, fractions d'histoire*. Basel, Boston, Berlin: Birkhäuser, 1992, 39–52. (Caveing, "Egyptian quantieme")
Černý, Jaroslav, and Gardiner, Alan H. *Hieratic Ostraca*. Oxford: Griffith Institute, 1957. (Černý/Gardiner, *Hieratic ostraca*)
Chace, Arnold B. *The Rhind Mathematical Papyrus. Free Translation and Commentary With Selected Photographs, Transcriptions, Transliterations, and Literal Translations* (Classics in Mathematics Education 8). Reston, VA: The National Council of Teachers of Mathematics, 1979. (Chace, *Rhind Mathematical Papyrus*)
Chace, Arnold B., Bull, Ludlow, Manning, Henry P., and Archibald, Raymond C. *The Rhind Mathematical Papyrus: British Museum 10057 and 10058*. Oberlin, OH: Mathematical Association of America, 1927 and 1929. (Chace/Bull/Manning/Archibald, *Rhind Mathematical Papyrus*)
Chauveau, Michel. *Egypt in the Age of Cleopatra: History and Society under the Ptolemies* (Trans. David Lorton). Ithaca, NY: Cornell University Press, 2000. (Chauveau, *Egypt in the Age of Cleopatra*)
Christianidis, Jean (ed.). *Classics in the History of Greek Mathematics* (Boston Studies in the Philosophy of Science 240). Leiden: Brill, 2004. (Christianidis, *Classics*)
Clagett, Marshall. *Ancient Egyptian Science. A Source Book. Vol. 3: Ancient Egyptian Mathematics* (Memoirs of the American Philosophical Society 232). Philadelphia: American Philosophical Society, 1999. (Clagett, *Egyptian Mathematics*)
Collier, Mark, and Quirke, Stephen: *The UCL Lahun Papyri* (Vol. 1: Letters, Vol. 2: Religious, Literary, Legal, Mathematical and Medical, Vol.3: Accounts; British Archaeological Reports International Series 1083/1209/1471). Oxford: Archaeopress, 2002/2004/2006. (Collier/Quirke, *UCL Lahun Papyri*)
Cooper, Jerrold S. "Babylonian beginnings: The origin of the cuneiform writing system in comparative perspective." In Stephen D. Houston, *The First Writing: Script Invention as History and Process*. Cambridge: Cambridge University Press, 2004, 71–99. (Cooper, "Babylonian beginnings")
Couchoud, Sylvia. *Mathématiques égyptiennes*. Paris: Editions Le Léopard d'Or, 1993. (Couchoud, *Mathématiques égyptiennes*)
——. "Le nombre pi et les surfaces rondes dans l'´Egypte pharaonique du Moyen Empire." *Studien zur Altägyptischen Kultur* 14 (1987): 35–42 (Couchoud: "Surfaces rondes")
Cour-Marty, Marguerite-Annie: "Les poids égyptiens, de precieux jalons archéologiques." *Cahier de recherches de l'Institut de Papyrologie et d'Egyptologie de Lille* 12 (1990): 17–55. (Cour-Marty, "Poids égyptiens")
Couyat, J., and Montet, Pierre: *Les inscriptions hiéroglyphiques et hiératiques du Ouâdi Hammâmât* (Mémoires publiés par les membres de l'Institut français d'archéologie orientale du Caire). Cairo: Institut français d'archéologie orientale du Caire. 1912. (Couyat/Montet, *Ouâdi Hammâmât*)
Cuomo, Serafina. *Ancient Mathematics*. London: Routledge, 2001. (Cuomo, *Ancient Mathematics*)
Curran, Brian A., Grafton, Anthony, Long, Pamela O., and Weiss, Benjamin: *Obelisk: A History*. Cambridge, MA: MIT Press, 2009. (Curran/Grafton/Long/Weiss, *Obelisk*)
Daressy, Georges.. "Calculs égyptiens du Moyen-Empire," *Recueil de Travaux* 28 (1906): 62–72. (Daressy, "Calculs égyptiens")
——. *Catalogue général des antiquités égyptiennes du Musée du Caire: Nos 25001–25385: Ostraca*. Cairo: Institut français d'archéologie orientale du Caire, 1901. (Daressy, *Catalogue général*)
——. "Un edifice archaique a Nazlet Batran." *Annales du service des antiquités de l'Égypte* 6 (1905): 99–106. (Daressy, "Edifice")
Davies, Benedict G. *Egyptian Historical Records of the Later Eighteenth Dynasty. Fascicle IV*. Warminster: Aris and Phillips Ltd., 1992. (Davies, *Historical records*)

Davies, Norman de Garis. *The Rock Tombs of Deir el Gebrâwi* (Archaeological Survey of Egypt 11). London: Egypt Exploration Society, 1902. (Davies, *Deir el Gebrâwi*)

Davies, Norman de Garis. *The Tomb of Antefoker, Vizier of Sesostris I, and of his Wife Senet* (No. 60; Theban Tomb Series 2). London: Egypt Exploration Fund, 1920. (Davies, *Antefoker*)

———. *The Tomb of Rekh-mi-Rē' at Thebes*. 2 vols. New York: Plantin Press, 1944. (reprinted 1973) (Davies, *Tomb of Rekh-mi-Rē'*)

———. *The Tombs of Two Officials of Tuthmosis the Fourth* (Nos. 75 and 90; Theban Tomb Series 3). London: Egypt Exploration Society, 1923. (Davies, *Tombs of two officials*)

Davis, Whitney. *Masking the Blow. The Scene of Representation in Late Prehistoric Egyptian Art*. Berkeley: University of California Press, 1992.

Depuydt, Leo. *Civil Calendar and Lunar Calendar in Ancient Egypt*. Leuven: Peeters en Departement Oosterse Studies, 1997. (Depuydt, *Civil calendar*)

Dibner, Bern. *Moving the Obelisks: A Chapter in Engineering History in which the Vatican Obelisk in Rome in 1586 Was Moved by Muscle Power, and a Study of More Recent Similar Moves*. Cambridge, MA: MIT Press, 1970 (Dibner, *Moving the obelisks*)

Donker van Heel, K.: *The legal manual of Hermopolis [pMattha], text and translation* (Uitgaven van wege de stichting "Hets Leids Papyrologisch Instituut" 11). Leiden: Hets Leids Papyrologisch Instituut 1990 (Donker van Heel, *Legal manual of Hermopolis*)

Dorman, Peter. "Rekhmire." In Donald Redford, ed., *The Oxford Encyclopedia of Ancient Egypt*. Oxford: Oxford University Press, 2001, 131–32. (Dorman, "Rekhmire")

Dreyer, Günter. "The tombs of the first and second dynasties at Abydos and Saqqara." In Zahi Hawass, ed., *Pyramids—Treasures, Mysteries and New Discoveries in Egypt*. Vercelli: White Star, 2007, 76–93. (Dreyer, "The tombs of the first and second dynasties")

———. *Umm el-Qaab I. Das prädynastische Königsgrab U-j und seine frühen Schriftzeugnisse* (Archäologische Veröffentlichungen 86). Mainz: von Zabern, 1998. (Dreyer, *Umm el-Qaab I*)

———. *Umm el-Qaab II. Importkeramik aus dem Friedhof U in Abydos (Umm el-Qaab) und die Beziehungen Ägyptens zu Vorderasien im 4. Jahrtausend v. Chr.* (Archäologische Veröffentlichungen 92). Mainz: von Zabern, 2001. (Dreyer, *Umm el-Qaab II*)

Dziobek, Eberhard. *Die Gräber des Vezirs User-Amin: Theben Nr. 61 und 131* (Archäologische Veröffentlichungen 84). Mainz: von Zabern, 1994. (Dziobek, *Gräber des Vezirs User-Amin*)

———. "Theban tombs as a source for historical and biographical evaluation: The case of User-Amun." In Jan Assmann, Eberhard Dziobek, Heike Guksch, and Friederike Kampp, eds., *Thebanische Beamtennekropolen. Neue Perspektiven archäologischer Forschung*. Internationales Symposion Heidelberg 9–13.6.1993 (Studien zur Archäologie und Geschichte Altägyptens 12). Heidelberg: Heidelberger Orientverlag, 1995, 129–40. (Dziobek, "Theban tombs")

Eisenlohr, August. *Ein mathematisches Handbuch der alten Ägypter*. Leipzig: Hinrichs, 1877. (Eisenlohr, *Mathematisches Handbuch*)

Ellis, Simon P. *Graeco-Roman Egypt*. Aylesbury: Shire Publications, 1992. (Ellis, *Graeco-Roman Egypt*)

Emery, Walter Bryan. *Excavations at Saqqara. The Tomb of Hemaka*. Cairo: Government Press, 1938. (Emery, *Hemaka*)

Engelbach, Reginald. *The Problem of the Obelisks, from a Study of the Unfinished Obelisk at Aswan*. London: T. F. Unwin, 1923. (Engelbach, *Problem of the obelisks*)

Engels, Hermann. "Quadrature of the circle in ancient Egypt." *Historia Mathematica* 4 (1977): 137–40. (Engels, "Quadrature of the circle")

Englund, Robert. *The State of Decipherment of Proto-Elamite* (Max Planck Institute for the History of Science Preprint 183). Berlin: Max Planck Institute for the History of Science, 2001. (Englund, *Decipherment*)

Epron, Lucienne, Daumas, François, and Goyon, George: *Le tombeau de Ti*. Vol. 1 (MIFAO 65). Cairo: Institut Français d'Archéologie Orientale, 1939. (Epron/Daumas/Goyon, *Ti*)

Erichsen, W. *Papyrus Harris I. Hieroglyphische Transkription*. (Bibliotheca Aegyptiaca 5) Bruxelles: Fondation Égyptologique Reine Élisabeth, 1933. (Erichsen, *Papyrus Harris I*)

Erman, Adolf. *Die altägyptischen Schülerhandschriften* (Abhandlungen der Preussischen Akademie der Wissenschaften. Jahrgang 1925. Phil.-Hist. Klasse, Nr. 2). Leipzig: Verlag der Akademie der Wissenschaften, 1925. (Erman, *Altägyptischen Schülerhandschriften*)

Eyre, Christopher J. "Weni's career and Old Kingdom historiography." In Christopher J. Eyre, Anthony Leahy, and Lisa Montagno Leahy, eds., *The Unbroken Reed. Studies in the Culture and Heritage of Ancient Egypt in Honour of A.F. Shore*. London: Egypt Exploration Society, 1994, 107–24. (Eyre, "Weni's career")

―――. "Work and the organisation of work in the New Kingdom," In Marvin A. Powell, ed., *Labor in the Ancient Near East* (American Oriental Series 68). New Haven, CT: American Oriental Society, 1987, 167–221. (Eyre, "Work and organization of work")

Faltings, Dina. *Die Keramik der Lebensmittelproduktion im Alten Reich— Ikonographie und Archäologie eines Gebrauchsartikels* (Studien zur Archäologie und Geschichte Altägyptens 14). Heidelberg: Heidelberger Orientverlag, 1998. (Faltings, *Lebensmittelproduktion*)

Felber, Heinz. *Demotische Ackerpachtverträge der Ptolemäerzeit. Untersuchungen zu Aufbau, Entwicklung und inhaltlichen Aspekten einer Gruppe von demotischen Urkunden* (Ägyptologische Abhandlungen 58). Wiesbaden: Otto Harrassowitz, 1997. (Felber, *Demotische Ackerpachtverträge*)

Fischer-Elfert, Hans-Werner "Berufliche Bildung und Ausbildung im alten Ägypten." In Max Liedtke, ed., *Berufliche Bildung—Geschichte, Gegenwart, Zukunft*. Bad Heilbrunn: Klinkhardt, 1997, 27–52. (Fischer-Elfert, "Ausbildung")

―――. *Die satirische Streitschrift des Papyrus Anastasi I. Textzusammenstellung* (Kleine ägyptische Texte). Wiesbaden: Otto Harrassowitz, 1983. (Fischer-Elfert, *Anastasi I-Text*)

―――. *Die satirische Streitschrift des Papyrus Anastasi I. Übersetzung und Kommentar* (Ägyptologische Abhandlungen 44). Wiesbaden: Otto Harrassowitz, 1986. (Fischer-Elfert, *Anastasi I-Übersetzung*)

―――. "Der Schreiber als Lehrer in der frühen ägyptischen Hochkultur." In Johann Georg Prinz von Hohenzollern and Max Liedtke, eds., *Schreiber, Magister, Lehrer. Zur Geschichte und Funktion eines Berufsstandes*. Bad Heilbrunn: Klinkhardt, 1989, 60–70 (Fischer-Elfert, "Schreiber als Lehrer")

Fletcher, E.N.R. "The area of the curved surface of a hemisphere in ancient Egypt." *Mathematical Gazette* 54 (1970): 227–29. (Fletcher, "Hemisphere")

Fowler, David. *The Mathematics of Plato's Academy. A New Reconstruction.* Oxford: Oxford University Press, 1999. (Fowler, *Mathematics of Plato's Academy*)

Franke, Detlef. "Middle Kingdom." In Donald B. Redford. ed., *The Oxford Encyclopedia of Ancient Egypt*, Vol. 2. Oxford: Oxford University Press, 2001, 393–400. (Franke, "Middle Kingdom")

Fried, Michael N., and Unguru, Sabetai. *Apollonius of Perga's Conica: Text, Context, Subtext.* Leiden: Brill, 2001. (Fried/Unguru, *Conica*)

Garcia, Juan Carlos Moreno (ed.). *Ancient Egyptian Administration*. Leiden: Brill 2013. (Garcia, *Administration*)

Gardiner, Alan H. *Ancient Egyptian onomastica.* 3 Vols. Oxford: Oxford University Press, 1947 (Gardiner, *Onomastica*)

―――. "The autobiography of Rekhmerēꜥ." *Zeitschrift für ägyptische Sprache* 60 (1925): 62–76. (Gardiner, "Autobiography of Rekhmerēꜥ")

―――. "The daily income of Sesostris II's Funerary Temple," *Journal of Egyptian Archaeology* 42 (1956): 119 (Gardiner, "Daily income")

―――. *Egyptian grammar. Being an introduction to the study of hieroglyphs.* Third revised edition. Oxford: Oxford University Press 1957 (Gardiner, *Egyptian grammar*)

―――. *Egyptian Hieratic Texts—Series I: Literary Texts of the New Kingdom, Part I.* Leipzig: Hinrichs, 1911. (Gardiner, *Egyptian hieratic texts*)

―――. *Ramesside administrative documents.* Oxford: Oxford University Press 1948 (Gardiner, *Ramesside administrative documents*)

―――. "Regnal years and civil calendar in Pharaonic Egypt," *Journal of Egyptian Archaeology* 31 (1945): 11–28 (Gardiner, "Regnal years")

―――. *The Wilbour papyrus.* 4 vols. London: Oxford University Press 1941–1952 (Gardiner, *Wilbour papyrus*)

Gasse, Annie: *Données nouvelles administratives et sacerdotales sur l'organisation du domaine d'Amon, XXe–XXIe dynasties: à la lumière des papyrus Prachov, Reinhardt et Grundbuch* (Bibliothèque d'étude 104). Cairo: Institut français d'archéologie orientale du Caire 1988 (Gasse, *Données nouvelles*)

Gerdes, Paulus: "Three alternate methods of obtaining the ancient Egyptian formula for the area of a circle." *Historia Mathematica* 12 (1985): 261–68. (Gerdes, "Three Alternate Methods")

Gillings, Richard J. "The division of 2 by the odd numbers 3 to 101 from the recto of the Rhind mathematical papyrus (B.M. 10058)." *Australian Journal of Science* 18 (1955): 43–49.

―――. "The volume of a truncated pyramid in ancient Egyptian papyri." *The Mathematics Teacher* 57 (1964): 552–55. (Gillings, "Volume of a truncated pyramid")

―――. "The area of the curved surface of a hemisphere in ancient Egypt." *The Australian Journal of Science* 30 (1967): 113–16. (Gillings, "Hemisphere")

―――. *Mathematics in the Time of the Pharaohs.* Cambridge, MA: MIT Press, 1972. (Gillings, *Mathematics in the time of the pharaohs*.)

———. "Response to 'Some comments on R. J. Gillings' analysis of the 2/*n* table in the Rhind papyrus." *Historia Mathematica* 5 (1978): 221–27. (Gillings, "Response")

Gillings, Richard J., and Rigg, W.J.A. "The area of a circle in ancient Egypt." *Australian Journal of Science* 32 (1969–70): 197–99. (Gillings/Rigg, "Area of a circle")

Glanville, Stephen R. K. "The mathematical leather roll in the British Museum." *Journal of Egyptian Archaeology* 13 (1927): 232–39. (Glanville, "Mathematical Leather Roll")

———. "Working Plan for a Shrine." *Journal of Egyptian Archaeology* 16 (1930): 237–39. (Glanville, "Working plan of a shrine")

Godley, Alfred Dennis. *Herodotus: Histories.* English translation by A. D. Godley, 4 vols. Cambridge, MA: Harvard University Press, 1926. (Godley, *Herodotus*)

Godron, G. "Quel est le lieu de provenance de la 'Pierre de Palerme'?" *Chronique d'Egypte* 27 (1952): 17–22. (Godron, "Pierre de Palerme")

Gödecken, Karin Barbara. *Eine Betrachtung der Inschriften des Meten im Rahmen der sozialen und rechtlichen Stellung von Privatleuten im ägyptischen Alten Reich* (Ägyptologische Abhandlungen 29). Wiesbaden: Otto Harrassowitz, 1976. (Gödecken, *Meten*)

Goyon, Georges. *Nouvelles inscriptions rupestres du Wadi Hammamat.* Paris: Librairie d'Amérique et d'Orient Adrien-Maisonneuve, 1957. (Goyon, *Wadi Hammamat*)

Grajetzki, Wolfram. *Court Officials of the Egyptian Middle Kingdom.* London: Duckworth, 2009. (Grajetzki, *Court Officials*)

———. *The Middle Kingdom of Ancient Egypt.* London: Duckworth, 2006. (Grajetzki, *Middle Kingdom*)

Grandet, Pierre. *Le Papyrus Harris I (BM 9999;* Bibliothèque d'étude 109). 2 vol. Cairo: Institut français d'archéologie orientale du Caire, 1994. (Grandet, *Papyrus Harris I*)

Griffith, Francis L. "Notes on Egyptian weights and measures." *Proceedings of the Society of Biblical Archaeology* 14 (1892): 403–50. (Griffith, "Weights and measures")

———. *The Petrie Papyri. Hieratic Papyri from Kahun and Gurob.* London: Quaritch, 1898. (Griffith, *Petrie Papyri*)

———. "The Rhind mathematical papyrus." *Proceedings of the Society of Biblical Archaeology* 16 (1894): 164–73. (Griffith, "Rhind Mathematical Papyrus")

Guillemot, Michel. "Les notations et les pratiques opératoires permettent-elles de parler de 'fractions égyptiennes'?" In P. Benoit K. Chemla, and J. Ritter, eds., *Histoire de fractions, fractions d'histoire.* Basel, Boston, Berlin: Birkhäuser, 1996: 53–69. (Guillemot, "Notations")

Gunn, Battiscomb, and Peet, Thomas E. "Four geometrical problems from the Moscow mathematical papyrus." *Journal of Egyptian Archaeology* 15 (1929): 167–85. (Gunn/Peet, "Four geometrical problems")

Habachi, Labib. *The Obelisks of Egypt: Skyscrapers of the Past.* New York: Scribner, 1977. (Habachi, *Obelisks*)

Habachi, Labib, and Vogel, Carola. *Die unsterblichen Obelisken Ägyptens.* Mainz: von Zabern, 2000. (Habachi/Vogel, *Obelisken*)

Hagen, Frederik. "Literature, transmission, and the late Egyptian miscellanies." In Rachael J. Dann, ed., *Current research in Egyptology 2004. Proceedings of the Fifth Annual Symposium.* Oxford: Oxbow Books, 2006. 84–99. (Hagen, "Late Egyptian Miscellanies")

Haring, Ben J. J. "A systematic bibliography on Deir el-Medina 1980–1990" In R. J. Demarée and A., eds., *Village Voices. Proceedings of the Symposium 'Texts from Deir el-Medîna and their interpretation,' Leiden, May 31–June 1, 1991* (CNWS Publications 13). Leiden: Nederlands Instituut voor het Nabije Oosten, 1992, 111–40. (Haring, "Bibliography Deir el-Medina")

Hart, George. *The Routledge Dictionary of Egyptian Gods and Goddesses.* Abingdon: Routledge, 2005. (Hart, *Gods and godesses*)

Hartwig, Melinda K. "The tomb of Menna." In Kent R. Weeks, ed., *Valley of the Kings.* Vercelli: White Star, 2001: 398–407. (Hartwig, "Tomb of Menna")

Hayes, William C. *Ostraka and Name Stones from the Tomb of Sen-Mut (No. 71) at Thebes.* New York: Metropolitan Museum of Art, 1942. (Hayes, *Senmut*)

———. *A Papyrus of the Late Middle Kingdom in the Brooklyn Museum, Pap. Brooklyn 35.1446.* New York: Brooklyn Museum, 1955.

Heisel, Joachim P. *Antike Bauzeichnungen.* Darmstadt: Wissenschaftliche Buchgesellschaft, 1993. (Heisel, *Antike Bauzeichnungen*)

Helck, Wolfgang. *Materialien zur Wirtschaftsgeschichte des Neuen Reiches.* 6 vols. Wiesbaden: Akademie der Wissenschaften und der Literatur. 1961–69. (Helck, *Materialien zur Wirtschaftsgeschichte*)

———. "Palermostein." In Wolfgang Helck and Eberhard Otto, eds., *Lexikon der Ägyptologie.* Vol. IV. Wiesbaden: Otto Harrassowitz, 1982: 652–54. (Helck, "Palermostein")

———. *Untersuchungen zu den Beamtentiteln des ägyptischen Alten Reiches* (Ägyptologische Forschungen 18). Glückstadt, Hamburg, New York: J.J. Augustin 1954 (Helck, *Beamtentitel*)
———. *Wirtschaftsgeschichte des Alten Ägypten im 3. und 2. Jahrtausend vor Chr.* Leiden and Köln: Brill, 1975. (Helck, *Wirtschaftsgeschichte*)
Hodel-Hoenes, Sigrid. *Life and Death in Ancient Egypt.* Ithaca and London: Cornell University Press, 2000. (Hodel-Hoenes, *Life and death*)
Hoffmann, Friedhelm. "Die Aufgabe 10 des Moskauer mathematischen Papyrus." *Zeitschrift für ägyptische Sprache und Altertumskunde* 123 (1996): 19–26. (Hoffmann, "Aufgabe 10")
———. "Measuring Egyptian statues." In John Steele and Annette Imhausen, eds., *Under One Sky. Astronomy and Mathematics in the Ancient Near East* (Alter Orient und Altes Testament 297). Münster: Ugarit, 2002, 109–19. (Hoffmann, "Measuring Egyptian statues")
Hoffmann, Friedhelm, and Quack, Joachim Friedrich. *Anthologie der demotischen Literatur.* Münster: LIT Verlag, 2007. (Hoffmann/Quack, *Anthologie der demotischen Literatur*)
Hofmann, Tobias. "Die Autobiographie des Uni aus Abydos." *Lingua Aegyptia* 10 (2002): 225–37 (Hofmann, "Autobiographie des Uni")
Holladay, John S., Jr. "Judaeans (and Phoenicians) in Egypt in the late seventh to sixth centuries BC." In Gary N. Knoppers and Antoine Hirsch, eds., *Egypt, Israel and the Ancient Mediterranean World. Studies in Honour of Donald B. Redford* (Probleme der Ägyptologie 20). Leiden: Brill, 2004, 405–37. (Holladay, "Judaeans (and Phoenicians) in Egypt")
Houston, Stephen D. (ed.). *The First Writing. Script Invention as History and Process.* Cambridge: Cambridge University Press, 2004. (Houston, *First writing*)
Høyrup, Jens. "Conceptual divergence—canons and taboos—and critique: Reflections on explanatory categories." *Historia Mathematica* 31 (2004): 129–47. (Høyrup, "Conceptual divergence")
———. *Lengths, Widths, Surfaces: A Portrait of Old Babylonian Algebra and Its Kin.* New York: Springer, 2002. (Høyrup, *Lengths, Widths, Surfaces*)
———. "Mathematics." In Hubert Cancik and Helmuth Schneider, eds., *Brill's New Pauly. Encyclopaedia of the Ancient World.* Vol. 8. Leiden and Boston: Brill, 2006, 461–69. (Høyrup,"Mathematics")
———. "Pythagorean 'rule' and 'theorem'—mirror of the relation between Babylonian and Greek mathematics." In Johannes Renger, ed., *Babylon: Focus mesopotamischer Geschichte, Wiege früher Gelehrsamkeit, Mythos in der Moderne.* Saarbrücken: Saarbrücker Druckerei und Verlag, 1999, 393–407. (Høyrup, "Pythagorean rule and theorem")
Imhausen, Annette: "Die ꜥḥꜥ-Aufgaben der ägyptischen mathematischen Texte und ihre Lösung(en)." In Caris-Beatrice Arnst, Ingelore Hafemann, and Angelika Lohwasser, eds., *Begegnungen. Antike Kulturen im Niltal. Festgabe für Erika Endesfelder, Karl-Heinz Priese, Walter Friedrich Reineke, Steffen Wenig.* Leipzig: Wodtke und Stegbauer, 2001, 213–20 (Imhausen, "ꜥḥꜥ-Aufgaben")
———. *Ägyptische Algorithmen. Eine Untersuchung zu den mittelägyptischen mathematischen Aufgabentexten* (Ägyptologische Abhandlungen 65). Wiesbaden: Harrassowitz, 2003. (Imhausen, *Algorithmen*)
———. "The algorithmic structure of the Egyptian mathematical problem texts." In John M. Steele and Annette Imhausen, eds., *Under One Sky: Astronomy and Mathematics in the Ancient Near East* (Alter Orient und Altes Testament 297). Münster: Ugarit-Verlag, 2002, 147–66. (Imhausen, "Algorithmic structure")
———. "Calculating the daily bread: Rations in theory and practice." *Historia Mathematica* 30 (2003): 3–16. (Imhausen, "Calculating the daily bread")
———. "Egyptian Mathematical Texts and Their Contexts." *Science in Context* 16 (2003): 367–89. (Imhausen, "Egyptian Mathematical Texts and Their Contexts")
———. "Egyptian mathematics," In Victor J. Katz, ed., *The Nathematics of Egypt, Mesopotamia, China, India, and Islam. A Sourcebook.* Princeton: Princeton University Press, 2007, 7–56 (Imhausen, "Egyptian mathematics")
———. "Mathematik und Mathematiker im pharaonischen Ägypten." *Mitteilungen der Mathematischen Gesellschaft Hamburg* 33 (2013): 75–97. (Imhausen, "Mathematik und Mathematiker")"
———. "Die Mathematisierung von Getreide im Alten Ägypten." *Mathematische Semesterberichte* 57 (2010): 3–10. (Imhausen, "Mathematisierung von Getreide")
———. "Traditions and myths in the historiography of Egyptian mathematics." In Eleanor Robson and Jacqueline Stedall, eds., *The Oxford Handbook of the History of Mathematics.* Oxford: Oxford University Press, 2009, 781–800 (Imhausen, "Myths")
———. "Normative structures in ancient Egyptian mathematical texts." In Daliah Bawanypeck and Annette Imhausen, eds., *Traditions of Written Knowledge in Ancient Egypt and Mesopotamia.* Münster: Ugarit Verlag, 2014, 153–87 (Imhausen, "Normative Structures")

———. "UC 32107A verso: A mathematical exercise?" In Mark Collier and Stephen Quirke, eds., *The UCL Lahun Papyri*. Vol. 3: *Accounts*. Oxford: Archaeopress, 2006, 288–301 (Imhausen, "UC 32107A")

Imhausen, Annette, and Ritter, Jim. "Mathematical fragments: UC 32114, UC 32118, UC 32134, UC 32159 UC32162." In Mark Collier and Stephen Quirke, eds., *The UCL Lahun Papyri*. Vol. 2: *Religious, Literary, Legal, Mathematical and Medical*. Oxford: Archaeopress, 2004, 71–96. (Imhausen/ Ritter, "Mathematical fragments")

Iversen, Erik. *Canon and Proportions in Egyptian Art*, 2nd rev. ed. Warminster: Aris and Phillips, 1975. (Iversen, *Canon and proportions*)

James, T.G.H. *The Hekanakhte Papers and Other Early Middle Kingdom Documents*. New York: Metropolitan Museum of Art, 1962. (James, *Hekanakhte*)

Janssen, Jac. J. "The role of the temple in the Egyptian economy during the New Kingdom." In Edward Lipinski, ed., *State and Temple Economy in the Ancient Near East: Proceedings of the International Conference Organized by the Katholieke Universiteit Leuven from the 10th to the 14th of April 1978* (Orientalia Lovaniensia Analecta 6), Vol. 2. Leuven: Departement Orientalistiek, 1979: 505–15. (Janssen, "Role of the temple")

Janssen, Jac. J. "Agrarian administration in Egypt during the Twentieth Dynasty." *Bibliotheca Orientalis* 43 (1986): 351–66. (Janssen, "Agrarian administration")

Jones, Dilwyn. *An Index of Ancient Egyptian Titles, Epithets and Phrases of the Old Kingdom*. 2 vols. Oxford: Archaeopress, 2000. (Jones, *Titles*)

Kahl, Jochem. "Die frühen Schriftzeugnisse aus dem Grab Uj in Umm el-Qaab." *Chronique d'Egypte* 78 (2003): 112–35. (Kahl, "Frühe Schriftzeugnisse")

———. *Das System der ägyptischen Hieroglyphenschrift in der 0.–3.Dynastie*. (Göttinger Orient Forschungen IV, Vol. 29). Wiesbaden: Harrassowitz, 1994. (Kahl, *Hieroglyphenschrift*)

Katary, Sally L.D. *Land Tenure in the Ramesside Period* (Studies in Egyptology). New York: Kegan Paul, 1989.

Katary, Sally L. D. "Land-tenure in the New Kingdom: The role of women smallholders and the military." In Alan K. Bowman and Eugene Rogan, *Agriculture in Egypt. From Pharaonic to Modern Times*. Oxford: Oxford University Press, 1999, 61–82. (Katary, "Land-Tenure")

———. "Taxation (until the end of the Third Intermediate Period." In Juan Carlos Moreno García and Willeke Wendrich, eds., *UCLA Encyclopedia of Egyptology*, Los Angeles, 2011. http://digital2 .library.ucla.edu/vievItem.do? ark=21198/zz002814vq. (Katary, "Taxation")

Katz, Victor. *A History of Mathematics. An Introduction*. New York: Harper Collins, 1993. (Katz, *History of mathematics*)

Kemp, Barry J. *Ancient Egypt. Anatomy of a Civilization*. New York: Routledge, 1989. (Kemp, *Ancient Egypt*)

Klebs, Luise. *Die Reliefs des Alten Reiches (2980–2475 v. Chr.). Material zur ägyptischen Kulturgeschichte*. Heidelberg: Winter, 1915. (Klebs, *Reliefs des Alten Reiches*)

———. *Die Reliefs und Malereien des Mittleren Reiches (VII.–XVII. Dynastie, ca. 2475–1580 v. Chr.). Material zur ägyptischen Kulturgeschichte*. Heidelberg: Winter, 1922. (Klebs, *Reliefs und Malereien des Mittleren Reiches*)

———. *Die Reliefs und Malereien des Neuen Reiches (XVIII.–XX. Dynastie, ca. 1580–1100 v. Chr.). Material zur ägyptischen Kulturgeschichte*. Heidelberg: Winter, 1934. (Klebs, *Reliefs und Malereien des Neuen Reiches*)

Kloth, Nicole. *Die (auto-)biographischen Inschriften des ägyptischen Alten Reiches: Untersuchungen zu Phraseologie und Entwicklung*. Hamburg: Helmut Buske Verlag, 2002. (Kloth, *Autobiographische Inschriften*)

Knorr, Wilbur R. "Techniques of fractions in ancient Egypt and Greece." *Historia Mathematica* 9 (1982): 133–71. (Knorr, "Techniques of fractions")

Kuhrt, Amélie. *The Ancient Near East, c. 3000–330 BC*. 2 vols. London/New York: Routledge, 1995. (Kuhrt, *Ancient Near East*)

Kurth, Dieter, and Behrmann, Almuth. *Die Inschriften des Tempels von Edfu*. Wiesbaden: Otto Harrassowitz, 2004. (Kurth/Behrmann, *Edfu*)

Lehner, Mark. *The Complete Pyramids*. London: Thames & Hudson, 1997. (Lehner, *Complete pyramids*)

Lepsius, Richard. *Die alt-aegyptische Elle und ihre Eintheilung* (Abhandlungen der königlichen Akademie der Wissenschaften). Berlin: Druckerei der königlichen Akademie der Wissenschaften, 1865. (Lepsius, *Alt-aegyptische Elle*)

Lesko, Leonard H. *Pharaoh's Workers: The Villagers of Deir el Medina*. Ithaca and London: Cornell University Press, 1994. (Lesko, *Pharaoh's workers*)

Lichtheim, Miriam: *Ancient Egyptian Literature. A Book of Readings. Volume 1: The Old and Middle Kingdoms.* Berkeley, Los Angeles, London: University of California Press, 2006 (1973). (Lichtheim, *Literature I*)

——. *Ancient Egyptian Literature. A Book of Readings. Volume 2: The New Kingdom.* Berkeley, Los Angeles, London: University of California Press, 2006 (1976). (Lichtheim, *Literature II*)

——. *Ancient Egyptian Literature. A Book of Readings. Volume 3: The Late Period.* Berkeley, Los Angeles, London: University of California Press, 2006 (1980). (Lichtheim, *Literature III*)

——. *Ancient Egyptian Autobiographies Chiefly of the Middle Kingdom. A Study and an Anthology* (Oribis Biblicus et Orientalis 84). Freiburg: Universitätsverlag Freiburg Schweiz, 1988. (Lichtheim, *Autobiographies*)

Lippert, Sandra Luisa. *Ein demotisches juristisches Lehrbuch. Untersuchungen zu Papyrus Berlin P23757rto* (Ägyptologische Abhandlungen 66). Wiesbaden: Harrassowitz, 2004. (Lippert, *Demotisches juristisches Lehrbuch*)

Lieven, Alexandra von. "Book of the dead, book of the living: BD spells as temple texts." *Journal of Egyptian Archaeology* 98 (2012): 249–67. (von Lieven, "Book of the dead, Book of the Living")

——. "Fragmente eines Feldregisters im Ashmolean Museum." *Studien zur Altägyptischen Kultur* 27 (1999): 255–60. (von Lieven, "Fragmente eines Feldregisters")

Lloyd, Alan B. *Herodotus Book II: Introduction and Commentary.* 3 volumes. Leiden: Brill, 1979–1988. (Lloyd, *Herodotus Book II*)

——. "The Late Period (664–332 BC)." In Ian Shaw, ed., *The Oxford History of Ancient Egypt.* Oxford: Oxford University Press, 2000, 369–93. (Lloyd, "Late Period")

——. "The Ptolemaic Period (332–30 BC)." In Ian Shaw, ed., *The Oxford History of Ancient Egypt.* Oxford: Oxford University Press, 2000, 395–420. (Lloyd, "Ptolemaic Period")

Loffet, Henri. *Les scribes comptables, les mesureurs de cereals et de fruits, les métreurs-arpenteurs et les peseurs de l'Égypte ancienne (de l'époque thinite à la xxie dynastie).* Lille: Atelier national de reproduction des theses, 2001. (Loffet, *Scribes*)

López, Jesus. *Ostraca Ieratici N.57093-57319* (Catalogo del Museo Egizio di Torino, Serie Seconda-Collezioni, Vol. III, Fasciolo 2). Milan: Istituto Editoriale Cisalpino-La Goliardica, 1980. (López, *Ostraca Ieratici*)

Loprieno, Antonio. *Ancient Egyptian. A Linguistic Introduction.* Cambridge: Cambridge University Press, 1995. (Loprieno, *Ancient Egyptian*)

Luckey, Paul. "Der Rauminhalt des ägyptischen Pyramidenstumpfes." *Zeitschrift für mathematischen und naturwissenschaftlichen Unterricht aller Schulgattungen* 63 (1932): 389–91. (Luckey, "Rauminhalt")

Malek, Jaromir. "The Old Kingdom." In Ian Shaw, ed., *The Oxford History of Ancient Egypt.* Oxford: Oxford University Press, 2000, 89–117. (Malek, "Old Kingdom")

Malek, Jaromir, and Forman, Werner. *In the Shadow of the Pyramids: Egypt during the Old Kingdom.* London: Orbis, 1986. (Malek/Forman, *Shadow of the pyramids*)

Manniche, Lise. *An ancient Egyptian herbal.* London: British Museum Press, 1989. (Manniche, *Egyptian herbal*)

Manning, Joseph Gilbert. *Land and Power in Ptolemaic Egypt. The Structure of Land Tenure.* Cambridge: Cambridge University Press, 2007. (Manning, *Land and power*)

Mariette, Auguste. *Abydos.* Vol. 2. Paris: Imprimerie nationale, 1880. (Mariette, *Abydos*)

——. *Les papyrus égyptiens du Musée de Boulaq,* Tome 2. Paris: Librairie A. Franck, 1872. (Mariette, *Papyrus Boulaq*)

Martin, Karl. *Ein Garantsymbol des Lebens* (Hildesheimer Ägyptologische Beiträge 3). Hildesheim: Gerstenberg Verlag, 1977. (Martin, *Garantsymbol*)

Mattha, Girgis. *The Demotic Legal Code of Hermopolis West* (Preface, additional notes and glossary by George R. Hughes; Bibliothèque d'étude 45). Cairo: Institut Français d'Archéologie Orientale, 1975. (Mattha, *Demotic legal code*)

McDowell, Andrea G. *Village Life in Ancient Egypt: Laundry Lists and Love Songs.* Oxford: Clarendon, 2001. (McDowell, *Village life*)

Meeks, Dimitri. *Le grand texte des donations au Temple d'Edfou* (Bibliothèque d'étude 59). Cairo: Institut français d'archéologie orientale du Caire, 1972. (Meeks, *Texte des donations*)

Melville, Duncan. "Poles and walls in Mesopotamia and Egypt," *Historia Mathematica* 31 (2004): 148–162. (Melville, "Poles and walls")

Menninger, Karl. *Number Words and Number Symbols: A Cultural History of Numbers.* New York: Dover Publications, 1969. (Menninger, *Number words*)

Menu, Bernadette (ed.). *L'organisation du travail en Égypte ancienne et en Mésopotamie* (Bibliothèque d'étude 151). Cairo: Institut français d'archéologie orientale du Caire, 2010. (Menu, *Organisation du travail*)

——. *Le régime juridique des terres et du personnel attaché à la terre dans le papyrus Wilbour* (Publications de la Faculté des Lettres et Sciences Humaines de l'Université de Lille 17). Lille: P. Geuthner, 1970. (Menu, *Régime juridique des terres*)

Meskell, Lynn. "Intimate archaeologies: The case of Kha and Merit." *World Archaeology* 29 (1998): 363–379. (Meskell, "Intimate archaeologies")

——. *Private Life in New Kingdom Egypt.* Princeton: Princeton University Press, 2002. (Meskell, *Private life*)

Miatello, Luca. "The difference 5½ in a problem of rations from the Rhind mathematical papyrus." *Historia Mathematica* 35 (2008): 277–84. (Miatello, "A problem of rations")

——. "The *nb.t* in the Moscow mathematical papyrus and a tomb model from Beni Hassan." *Journal of Egyptian Archaeology* 96 (2010): 228–32. (Miatello, "Moscow Mathematical Papyrus and a tomb model")

——. "Problem 60 of the Rhind mathematical papyrus: Glaring errors or correct method?" *Journal of the American Research Center in Egypt* 45 (2009): 153–58. (Miatello, "Problem 60 of the Rhind Mathematical Papyrus")

Michel, Marianne. *Les mathématiques de l'Égypte ancienne. Numération, métrologie, arithmétique, géométrie et autres problèmes.* Bruxelles: Éditions Safran 2004.

Midant-Reynes, Béatrix. *The Prehistory of Egypt. From the First Egyptians to the First Pharaohs.* Oxford: Blackwell, 2000. (Midant-Reynes, *Prehistory*)

Möller, Astrid. *Naukratis: Trade in Archaic Greece* (Oxford Monographs on Classical Archaeology). Oxford: Oxford University Press, 2000. (Möller, *Naukratis*)

Möller, Georg. *Hieratische Paläographie. Die ägyptische Buchschrift in ihrer Entwicklung von der fünften Dynastie bis zur römischen Kaiserzeit.* 3 vols. Leipzig: J. C. Hinrichs, 1927. (Möller, *Paläographie*)

Moran, William L. *The Amarna Letters.* Baltimore and London: Johns Hopkins University Press, 1992. (Moran, *Amarna Letters*)

Morenz, Ludwig. "Der Fisch an der Angel. Die hieroglyphenbildliche Metapher eines Mathematikers." *Zeitschrift für ägyptische Sprache* 133 (2006): 51–55. (Morenz, "Fisch an der Angel")

——. "Die Systematisierung der ägyptischen Schrift im frühen 3. Jahrtausend v. Chr. Eine kultur- und schriftgeschichtliche Rekonstruktion," In Ludwig Morenz and R. Kuhn, ed., *Vorspann oder formative Phase? Ägypten und der Vordere Orient 3500–2700 v. Chr.* (Philippika 48). Wiesbaden: Harrassowitz, 2011, 19–47. (Morenz, "Systematisierung der ägyptischen Schrift")

Moussa, Ahmed M., and Altenmüller, Hartwig. *Das Grab des Nianchchnum und Chnumhotep* (Archäologische Veröffentlichungen 21). Mainz: von Zabern, 1977. (Moussa/Altenmüller, *Nianchchnum*)

Mueller, Dieter. "Some remarks on wage rates in the Middle Kingdom." *Journal of Near Eastern Studies* 34 (1975): 249–263. (Mueller, "Wage rates")

Muhs, Brian P. *Tax Receipts, Taxpayers and Taxes in Early Ptolemaic Thebes* (Oriental Institute Publications 126). Chicago: Oriental Institute of the University of Chicago, 2005. (Muhs, *Tax receipts*)

Murray, Mary Anne. "Cereal production and processing." In Paul T. Nicholson and Ian Shaw, eds., *Ancient Egyptian Materials and Technology.* Cambridge: Cambridge University Press, 2000, 505–36 (Murray: "Cereal production")

Naville, Edouard. *The Temple of Deir el Bahri. Part VI: The Lower Terrace, Additions and Plans* (Memoir of the Egypt Exploration Fund 29). London: Egypt Exploration Fund, 1908. (Naville, *Deir el Bahri VI*)

Neugebauer, Otto. "Zur ägyptischen Bruchrechnung." *Zeitschrift für ägyptische Sprache und Altertumskunde* 64 (1929): 44–48. (Neugebauer, "Lederrolle")

——. "Die Geometrie der ägyptischen mathematischen Texte." In Otto Neugebauer, Julius Stenzel, amd OttoToeplitz, eds., *Quellen und Studien zur Geschichte der Mathematik, Astronomie und Physik, Abteilung B. Studien*, Vol. 1. Berlin: Julius Springer, 1931, 413–51. (Neugebauer, "Geometrie")

——. *Die Grundlagen der ägyptischen Bruchrechnung.* New York: Springer, 1926. (Neugebauer, *Bruchrechnung*)

——. *A History of Ancient Mathematical Astronomy.* 3 vols. New York: Springer, 1975. (Neugebauer, *History of ancient mathematical astronomy*)

——. *Mathematische Keilschrift-Texte.* Vol. II. Berlin: Julius Springer, 1935. (Neugebauer, *Mathematische Keilschrift-Texte II*)

——. *Mathematische Keilschrift-Texte.* Vol. III. Berlin: Julius Springer, 1937. (Neugebauer, *Mathematische Keilschrift-Texte III*)

——. "Das Pyramidenstumpf-Volumen in der vorgriechischen Mathematik." In Otto Neugebauer, Julius Stenzel, and Otto Toeplitz, *Quellen und Studien zur Geschichte der Mathematik, Astronomie und Physik. Abteilung B. Studien*, Vol. 2. Berlin: Springer, 1933, 347–51. (Neugebauer, "Pyramidenstumpf-Volumen")

——. "Über die Konstruktion von *sp* „Mal" im mathematischen Papyrus Rhind." *Zeitschrift für ägyptische Sprache und Altertumskunde* 62 (1927): 61–62. (Neugebauer, "Konstruktion von sp")

——. *Vorlesungen über Geschichte der antiken mathematischen Wissenschaften. Erster Band: Vorgriechische Mathematik.* New York: Springer, 1969. (Neugebauer, *Vorlesungen*)

Newberry Percy E. *Beni Hasan: Part I* (Archaeological Survey of Egypt 1). London: Egypt Exploration Fund, 1893. (Newberry, *Beni Hassan I*)

——. *Beni Hasan: Part II* (Archaeological Survey of Egypt 2). London: Egypt Exploration Fund, 1894. (Newberry, *Beni Hassan II*)

Parker, Richard A. *The Calendars of Ancient Egypt* (Studies in Ancient Oriental Civilization 26). Chicago: University of Chicago Press, 1950. (Parker, *Calendars*)

——. "A demotic mathematical papyrus fragment." *Journal of Near Eastern Studies* 18 (1959): 275–79. (Parker, "Demotic mathematical fragment")

——. *Demotic Mathematical Papyri.* Providence, RI: Brown University Press, 1972. (Parker, *Demotic Mathematical Papyri*)

——. "A mathematical exercise: P. Dem. Heidelberg 663." *Journal of Egyptian Archaeology* 61 (1975): 189–96. (Parker, "Mathematical exercise")

Parkinson, Richard B. *The Painted Tomb-chapel of Nebamun.* London: British Museum Press, 2008. (Parkinson, *Nebamun*)

——. *Papyrus.* Austin, TX: University of Texas Press, 1995. (Parkinson, *Papyrus*)

——. *Poetry and Culture in Middle Kingdom Egypt. A Dark Side to Perfection.* London and New York: Continuum, 2002. (Parkinson, *Poetry and culture*)

——. *The Tale of Sinuhe and Other Ancient Egyptian Poems, 1940–1640 BC.* Oxford: Oxford University Press, 1997. (Parkinson, *Sinuhe*)

Peacock, David. "The Roman Period (30 BC–AD 395)." In Ian Shaw, ed., *The Oxford History of Ancient Egypt.* Oxford: Oxford University Press, 2000: 422–44. (Peacock, "Roman Period")

Peet, Thomas E. "Arithmetic in the Middle Kingdom," *Journal of Egyptian Archaeology* 9 (1923): 91–95. (Peet, "Arithmetic")

——. "Notices of recent publications: Mathematischer Papyrus des Staatlichen Museums der Schönen Künste in Moskau. Von W.W. STRUVE. (Quellen u. Studien zur Geschichte der Mathematik; Abteilung A: Quellen, Band I.) Berlin 1930," *Journal of Egyptian Archaeology* 17 (1931): 154–60. (Peet, "Review Struve")

——. "A problem in Egyptian geometry." *Journal of Egyptian Archaeology* 17 (1931): 100–6. (Peet, "Egyptian geometry")

——. *The Rhind Mathematical Papyrus, British Museum 10057 and 10058. Introduction, Transcription, Translation and Commentary.* London: Hodder and Stoughton, 1923. (Peet, *Rhind Mathematical Papyrus*)

Petrie, William Matthew Flinders. *Gizeh and Rifeh* (British School of Archaeology in Egypt 13). London: Bernard Quaritch, 1907. (Petrie, *Gizeh and Rifeh*)

Polz, Daniel. "An architects sketch from the Theban necropolis," *Mitteilungen des Deutschen Archäologischen Instituts Abteilung Kairo* 53 (1997): 233–40. (Polz, "An architects sketch")

Pommerening, Tanja. *Die altägyptischen Hohlmaße* (Studien zur Altägyptischen Kultur Beihefte 10). Hamburg: Helmut Buske, 2005. (Pommerening, *Hohlmaße*)

——. "Die *šsȝw*-Lehrtexte der heilkundlichen Literatur der Alten Ägypten. Tradition und Textgeschichte." In Daliah Bawanypeck and Annette Imhausen, eds., *Traditions of Written Knowledge in Ancient Egypt and Mesopotamia.* Münster: Ugarit Verlag, 2014, 7–46 (Pommerening, "Lehrtexte")

Porter, Bertha, and Moss, Rosalind L. B. *Topological Bibliography of Ancient Egyptian Hieroglyphic Texts, Reliefs and Paintings. Vol. IV: Lower and Middle Egypt.* Oxford: Clarendon Press, 1934. (Porter/Moss: *Topological Bibliography IV*)

Posener-Kriéger, Paule. *Les archives du temple funéraire de Néferirkarê-Kakaï: les papyrus d'Abousir. Traduction et commentaire* (Bibliothèque d'étude 65). 2 vols. Cairo: Institut français d'archéologie orientale du Caire, 1976 .(Posener-Kriéger, *Archives du temple*)

——. "Les mesures des étoffes à l'Ancien Empire," *Revue d'Egyptologie* 29 (1977): 86–96. (Posener-Kriéger, "Mesures des étoffes")

——. "Les papyrus de Gébelein. remarques préliminaires." *Revue d'Egyptologie* 27 (1975): 211–21. (Posener-Kriéger, "Papyrus de Gébelein")

Posener-Kriéger, Paule, and de Cenival, Jean-Louis. *The Abu Sir Papyri. Hieratic Papyri in the British Museum, Ser. 5.* London: British Museum Press, 1968. (Posener-Kriéger/de Cenival, *Hieratic papyri*)

Posener-Kriéger, Paule, and Demichelis, Sara. *I papiri di Gebelein.* Turin: Museo Egizio di Torino, 2004. (Posener-Kriéger/Demichelis, *Papiri di Gebelein*)

Quack, Joachim Friedrich. *Einführung in die altägyptische Literaturgeschichte III. Die demotische und gräko-ägyptische Literatur.* Münster: LIT Verlag, 2005. (Quack, *Altägyptische Literaturgeschichte III*)

Quibell, James Edward. *Archaic Objects* (Catalogue général des antiquités égyptiennes du Musée du Caire 23/24). 2 vols. Cairo: Imprimérie de l'Institut français d'archéologie orientale, 1904–5. (Quibell, *Archaic objects*)

———. *Hierakonpolis* (Egyptian Research Account, Fourth/Fifth Memoir). 2 vols. London: Bernard Quaritch, 1900/1902. (Quibell, *Hierakonpolis*)

Quirke, Stephen. *The Administration of Egypt in the Late Middle Kingdom: The Hieratic Documents.* New Malden: Sia Publishing, 1990. (Quirke, *Administration*)

———. "Archive." In Antonio Loprieno, ed. *Ancient Egyptian Literature: History and Forms.* Leiden: Brill, 1996, 379–401. (Quirke, "Archive")

———. *Titles and Bureaux of Egypt 1850–1700 BC.* London: Golden House Publications, 2004. (Quirke, *Titles*)

Rammant-Peeters, Agnes. *Les pyramidions égyptiens du nouvel empire* (Orientalia Lovaniensia Analecta 11). Leuven: Departement Orientalistiek, 1983. (Rammant-Peeters, *Pyramidions égyptiens*)

Rashed, Roshdi. *The Development of Arabic Mathematics: Between Arithmetic and Algebra.* Dordrecht, London: Kluwer, 1994. (Rashed, *Arabic mathematics*)

Rees, Charles S. "Egyptian fractions." *Mathematical Chronicle* 10 (1981): 13–30. (Rees, "Egyptian fractions")

Regulski, Ilona. "The origin of writing in relation to the emergence of the Egyptian state." In Béatrix Midant-Reynes and Yann Tristant, *Egypt at Its Origins 2. Proceedings of the International Conference "Origin of the State, Predynastic and Early Dynastic Egypt"; Toulouse (France), September 5–8, 2005* (Orientalia Lovaniensia analecta 172). Leuven: Peeters 2008: 985–1009. (Regulski, "Origin of writing")

Reineke, Walter-Friedrich. *Gedanken und Materialien zur Frühgeschichte der Mathematik in Ägypten.* London: Golden House Publications, 2014. (Reineke, *Gedanken und Materialien*)

Reineke, Walter-Friedrich. "Zur Ziegelrampe des Papyrus Anastasi I." *Altorientalische Forschungen* 2 (1975): 5–9. (Reineke, "Ziegelrampe")

Rising, Gerald R. "The Egyptian use of unit fractions for equitable distribution." *Historia Mathematica* 1 (1974): 93–94. (Rising, "Egyptian use of unit fractions")

Ritter Jim. "Chacun sa vérité: les mathématiques en Égypte et en Mésopotamie." In Michel Serres, ed., *Élements d'histoire des sciences.* Paris: Bordas, 1989, 39–61. (Ritter, "Mathématiques en Égypte et en Mésopotamie")

———. "Closing the eye of Horus," In John Steele and Annette Imhausen, eds., *Under One Sky. Astronomy and Mathematics in the Ancient Near East* (Alter Orient und Altes Testament 297). Münster: Ugarit, 2002, 297–323. (Ritter, "Eye of Horus")

———. "Egyptian mathematics." In Helaine Selin, ed., *Mathematics Across Cultures: The History of Non-Western Mathematics.* Dordrecht: Kluwer, 2000, 115–36. (Ritter, "Egyptian Mathematics")

———. "Mathematics in Egypt." In Helaine Selin, ed., *Encyclopedia of the History of Science, Technology, and Medecine in Non-Western Cultures.* Dordrecht, Boston, London: Kluwer, 1997, 629–32. (Ritter, "Mathematics in Egypt")

———. "Measure for measure: Mathematics in Egypt and Mesopotamia," In Michel Serres, ed., *A History of Scientific Thought. Elements of a History of Science.* Oxford and Cambridge, MA: Blackwell, 1995: 44–72. (Ritter, "Measure for measure")

———. "Metrology and the prehistory of fractions," In Paul Benoit, Karine Chemla, and Jim Ritter, eds., *Histoire des fractions, fractions d'histoire.* Basel, Boston, Berlin: Birkhäuser, 1992, 19–34. (Ritter, "Metrology")

———. "Reading Strasbourg 368: A thrice told tale." In Karine Chemla, ed., *History Of Science, History Of Text* (Boston Studies in the Philosophy of Science 238). New York: Springer, 2004, 177–200. (Ritter, "Reading Strasbourg 368")

Ritter, Jim, and Vitrac, Bernard. "La pensée orientale et la pensée grecque." In Jean François Mattei, ed., *L'Encyclopédie philosophique universelle. IV. Le discours philosophique.* Paris: Presses universitaires de France, 1998, 1233–50. (Ritter/Vitrac: "Pensée oriental")

Robins, Gay. "Canonical proportions and metrology," *Discussions in Egyptology* 32 (1995): 91–92. (Robins, "Canonical proportions and metrology")

———. *Proportion and Style in Ancient Egyptian Art.* London: Thames and Hudson, 1994. (Robins, *Proportion*)

Robins, Gay, and Shute, Charles. *The Rhind Mathematical Papyrus. An Ancient Egyptian Text.* London: British Museum Publications, 1987. (Robins/Shute, *Rhind Mathematical Papyrus*)

Robson, Eleanor. "Literacy, numeracy, and the state in early Mesopotamia." In: Kathryn Lomas, Ruth D. Whitehouse, and John B. Wilkins, eds., *Literacy and the State in the Ancient Mediterranean* (Specialist Studies on the Mediterranean 7). London: Accordia Research Institute, 2007, 37–50. (Robson, "Literacy")

———. *Mesopotamian Mathematics, 2100–1600 BC. Technical Constants in Bureaucracy and Education* (Oxford Editions of Cuneiform Texts). Oxford: Clarendon Press, 1999. (Robson, *Mesopotamian Mathematics*)

———. "More than metrology: Mathematics education in an Old Babylonian scribal school," In John Steele and Annette Imhausen, eds., *Under One Sky: Mathematics and Astronomy in the Ancient Near East* (Alter Orient und Altes Testament 297). Münster: Ugarit-Verlag, 2002, 325–65. (Robson, "More than metrology")

———. "Neither Sherlock Holmes nor Babylon: A reassessment of Plimpton 322," *Historia Mathematica* 28 (2001): 167–206. (Robson, "Plimpton 322")

———. "Tables and tabular formatting in Sumer, Babylonia, and Assyria, 2500–50 BCE." In M. Campbell-Kelly, M. Croarken, R. G. Flood, and Eleanor Robson, eds., *The History of Mathematical Tables from Sumer to Spreadsheets*. Oxford: Oxford University Press, 2003, 18–47. (Robson, "Tables")

Rochholz, Matthias.. *Schöpfung, Feindvernichtung, Regeneration: Untersuchung zum Symbolgehalt der machtgeladenen Zahl 7 im alten Ägypten*. Wiesbaden: Otto Harrassowitz, 2002. (Rochholz, *Symbolgehalt*)

Römer, Malte. *Gottes- und Priesterherrschaft in Ägypten am Ende des Neuen Reiches* (Ägypten und Altes Testament 21). Wiesbaden: Harrassowitz, 1994. (Römer, *Gottes- und Priesterherrschaft*)

Roik, Elke, *Das Längenmaßsystem im alten Ägypten*. Hamburg: Christian-Rosenkreutz, 1993. (Roik, *Längenmaßsystem*)

Rossi, Corina. *Architecture and Mathematics in Ancient Egypt*. Cambridge: Cambridge University Press, 2004. (Rossi, *Architecture and mathematics*)

Rossi, Corianna, and Imhausen, Annette. "Architecture and mathematics in the time of Senusret I: Sections G, H and I of papyrus Reisner I," In Salima Ikram and Aidan Dodson, eds., *Beyond the Horizon: Studies in Egyptian Art, Archaeology and History in Honour of Barry J. Kemp*. Cairo: Supreme Council of Antiquities Press, 2009, 440–55. (Rossi/Imhausen, "Papyrus Reisner I")

Schack-Schackenburg, Hans. "Der Berliner papyrus 6619 (mit 1 Tafel)." *Zeitschrift für Ägyptische Sprache und Altertumskunde* 38 (1900): 135–40. (Schack-Schackenburg, "Berlin Papyrus 6619")

———. "Das kleiner Fragment des Berliner Papyrus 6619." *Zeitschrift für Ägyptische Sprache und Altertumskunde* 40 (1902): 65–66. (Schack-Schackenburg, "Kleineres Fragment")

Schaedel, Herbert D. Die Listen des großen Papyrus Harris. Ihre wirtschaftliche und politische Ausdeutung. Glückstadt/Hamburg/New York: J. J. Augustin, 1936. (Schaedel, *Listen*)

Scharff, Alexander. "Ein Rechnungsbuch des königlichen Hofes aus der 13. Dynastie." *Zeitschrift für ägyptische Sprache* 57 (1922): 51–68 and 1**24**. (Scharff, "Rechnungsbuch")

Schlott-Schwab, Adelheid, *Die Ausmasse Ägyptens nach altägyptischen Texten* (Ägypten und Altes Testament 3). Wiesbaden: Harrassowitz. 1981. (Schlott-Schwab, *Ausmasse*)

Schott, Siegfried. *Untersuchungen zum Ursprung der Schrift* (Akademie der Wissenschaften und der Literatur in Mainz: Abhandlungen der Geistes- und Sozialwissenschaftlichen Klasse, Jahrgang 1950, Nr. 24). Wiesbaden: Franz Steiner, 1950. (Schott, *Ursprung der Schrift*)

Scott, A., and Hall, H. R. "Laboratory notes: Egyptian leather roll of the seventeenth century BC," *The British Museum Quarterly* 2 (1927): 56–57. (Scott/Hall, "Laboratory Notes")

Scott, Nora E. "Egyptian cubit rods." *Bulletin of the Metropolitan Museum of Arts* N.S. 1 (1942): 70–75. (Scott, "Egyptian Cubit Rods")

Seeber, Christine. *Untersuchungen zur Darstellung des Totengerichts im Alten Ägypten* (Münchner Ägyptologische Studien 35). München, Berlin: Deutscher Kunstverlag, 1976. (Seeber, *Totengericht*)

Seidlmayer, Stephan. "Computer im Alten Ägypten. Aus der Urgeschichte der Datenverarbeitung." *Gegenworte* 8 (2001): 69–71. (Seidlmayer, "Computer im Alten Ägypten")

———. "The first intermediate period (ca. 2160–2055 BC)." In Ian Shaw, ed., *The Oxford History of Ancient Egypt*. Oxford: Oxford University Press, 2000, 118–47 (Seidlmayer, "First Intermediate Period")

———. *Historische und moderne Nilstände. Untersuchungen zu den Pegelablesungen des Nils von der Frühzeit bis in die Gegenwart*. Berlin: Achet Verlag, 2001. (Seidlmayer, *Nilstände*)

Sesiano, Jacques. "Survivance médiévale en Hispanie d'un probleme né en Mésopotamie." *Centaurus* 30 (1987): 18–61. (Sesiano, "Survivance")

Sethe, Kurt. *Die Einsetzung des Veziers unter der 18. Dynastie* (Untersuchungen zur Geschichte und Altertumskunde Aegyptens 5-2). Leipzig: J. C. Hinrichs, 1909. (Sethe, *Einsetzung des Veziers*)

———. *Urkunden des ägyptischen Altertums. Abteilung I: Urkunden des Alten Reiches*, 2nd ed. Leipzig: Hinrichs, 1932. (Sethe, *Urkunden des Alten Reichs*)

———. *Von Zahlen und Zahlworten bei den alten Ägyptern und was für andere Völker und Sprachen daraus zu lernen ist. Ein Beitrag zur Geschichte von Rechenkunst und Sprache*. Straßburg: Karl J. Trübner, 1916. (Sethe, *Zahlen und Zahlworte*)

Shaw, Ian (ed.). *The Oxford History of Ancient Egypt*. Oxford: Oxford University Press, 2000. (Shaw, *History*)

Shirley, Judith J. "Viceroys, viziers & the Amun precinct: The power of heredity and strategic marriage in the early 18th dynasty." *Journal of Egyptian History* 3 (2010): 73–113. (Shirley, "Viceroys, viziers & the Amun precinct")

Shute, Charles. "Mathematics." In Donald B. Redford, ed., *The Oxford Encyclopedia of Ancient Egypt*. Oxford: Oxford University Press, 2001: 348–51. (Shute, "Mathematics")

Sigler, Laurence E. (trans.). *Fibonacci's Liber Abaci*. New York: Springer, 2002. (Sigler, Fibonacci-Liber Abaci)

Simpson, William Kelly. *The Literature of Ancient Egypt. An Anthology of Stories, Instructions, Stelae, Autobiographies, and Poetry*. New Haven and London: Yale University Press, 2003. (Simpson, *Literature*)

———. *Papyrus Reisner I. The Records of a Building Project in the Reign of Sesostris I*. Boston: Museum of Fine Arts, 1963. (Simpson, *Reisner I*)

———. *Papyrus Reisner II. Accounts of the Dockyard Workshop at This in the Reign of Sesostris I*. Boston: Museum of Fine Arts, 1965. (Simpson, *Reisner II*)

———. *Papyrus Reisner III. The Records of a Building Project in the Early Twelfth Dynasty*. Boston: Museum of Fine Arts, 1969. (Simpson, *Reisner III*)

———. *Papyrus Reisner IV. Personnel Accounts of the Early Twelfth Dynasty*. Boston: Museum of Fine Arts, 1986. (Simpson, *Reisner IV*)

Smeur, A.J.E.M. "On the value equivalent to π in ancient mathematical texts: A new interpretation." *Archive for History of Exact Sciences* 6 (1970): 249–70. (Smeur, "Value equivalent to π")

Smither, Paul. "A tax-assessor's journal of the Middle Kingdom." *Journal of Egyptian Archaeology* 27 (1941): 74–76. (Smither, "Tax-assessor's journal")

Spalinger, Anthony. "Assurbanipal and Egypt: A Source Study." *Journal of the American Oriental Society* 94 (1974): 316–28. (Spalinger, "Assurbanipal and Egypt")

———. "Baking during the reign of Seti I." *Bulletin de l'Institut Français d'Archéologie Orientale du Caire* 86 (1986): 307–52. (Spalinger, "Baking during the reign of Seti I")

———. "Dates in ancient Egypt." *Studien zur Altägyptischen Kultur* 15 (1988): 255–76. (Spalinger, "Dates in ancient Egypt")

———. "Review of Marshall Clagett: Ancient Egyptian science. A source book. Vol. 3, Ancient Egyptian Mathematics (Memoirs of the American Philosophical Society 232). Philadelphia: American Philosophical Society 1999." *Journal of the American Oriental Society* 121 (2001): 133. (Spalinger, "Review Clagett")

———. *Revolutions in Time: Studies in Ancient Egyptian Calendrics*. San Antonio: Van Siclen Books, 1994. (Spalinger, *Revolutions in time*)

———. "The Rhind mathematical papyrus as a historical document." *Studien zur Altägyptischen Kultur* 17 (1990): 295–337. (Spalinger, "Rhind Mathematical Papyrus as a historical document")

Spencer, Jeffrey (ed.). *The British Museum Book of Ancient Egypt*. London: British Museum Press, 2007. (Spencer, *Ancient Egypt*)

Spencer, Patricia. *The Egyptian Temple. A Lexicographical Study*. London, Boston, Melbourne and Henley: Kegan Paul International, 1984. (Spencer, *Temple*)

Spiegelberg, Wilhelm. Rechnungen aus der Zeit Setis I. Strassburg: Trübner, 1896. (Spiegelberg, *Rechnungen*)

Stadler, Martin. "Judgment after death (negative confession)." In Jacco Dieleman and Willeke Wendrich, eds., *UCLA Encyclopedia of Egyptology*. Los Angeles, 2008. http://escholarship.org/uc/item/07 s1t6kj. (Accessed August 5, 2013.) (Stadler, "Judgment after death")

Steindorff, Georg. *Das Grab des Ti in 143 Lichtdrucktafeln und 20 Blättern*. Leipzig: J. C. Hinrichs'sche Buchhandlung, 1913. (Steindorff, *Grab des Ti*)

Strudwick, Nigel. *The Administration of Egypt in the Old Kingdom: The Highest Titles and Their Holders*. London: KPI, 1985. (Strudwick, *Administration*)

———. *Texts from the Pyramid Age* (Writings from the ancient world 16). Atlanta: Society of Biblical Literature, 2005. (Strudwick, *Pyramid age*)

Struve, Wasili W. *Mathematischer Papyrus des staatlichen Museums der schönen Künste in Moskau* (Quellen und Studien zur Geschichte der Mathematik, Astronomie und Physik, Abteilung A: Quellen 1). Berlin: Julius Springer, 1930. (Struve, *Mathematischer Papyrus Moskau*)

Tacke, Nikolaus. *Verspunkte als Gliederungsmittel in ramessidischen Schülerhandschriften* (Studien zur Archäologie und Geschichte Altägyptens 22). Heidelberg: Heidelberger Orientverlag, 2001. (Tacke, *Verspunkte*)

Tait, W. John. "Demotic literature: Forms and genres," In Antonio Loprieno, ed., *Ancient Egyptian Literature: History and Forms* (Probleme der Ägyptologie 10). Leiden: Brill, 1996,175–87. (Tait, "Demotic literature")

Taylor, John H. *Death and the Afterlife in Ancient Egypt*. London: British Museum Press, 2001.

Thissen, Heinz Josef. "Die demotische Literatur als Medium spätägyptischer Geisteshaltung." In Günter Burkard, Alfred Grimm, Sylvia Schoske, and Alexandra Verbovsek, eds., *Kon-Texte. Akten des Symposions "Spurensuche—Altägypten im Spiegel seiner Texte,"* München 2. bis 4. Mai 2003 (Ägypten und Altes Testament 60). Wiesbaden: Harrassowitz, 2003, 91–101. (Thissen, "Demotische Literatur")

Thomas, W. R. "Moscow mathematical papyrus, no. 14," *Journal of Egyptian Archaeology* 17 (1931): 50–52. (Thomas, "Moscow mathematical papyrus, no. 14")

Unguru, Sabetai. "On the need to rewrite the history of Greek mathematics," *Archive for History of Exact Sciences* 15 (1976): 67–114. (Unguru, "Need to Rewrite")

Unguru, Sabetai, and Rowe, David. "Does the quadratic equation have Greek roots? A study of 'geometric algebra,' 'application of areas,' and related problems (part 1)." *Libertas Mathematica* 1 (1981): 1–49. (Unguru/Rowe, "Quadratic Equation I")

———. "Does the quadratic equation have Greek roots? A study of 'geometric algebra,' 'application of areas,' and related problems (part 2)," *Libertas Mathematica* 2 (1982): 1–62. (Unguru/Rowe, "Quadratic Equation II")

Urton, Gary. *The Social Life of Numbers: A Quechua Ontology of Numbers and Philosophy of Arithmetic*. Austin: University of Texas Press, 1997. (Urton, *Social life of numbers*)

van den Boorn, G.P.F. *The Duties of the Vizier*. London: Kegan Paul International, 1988. (van den Boorn, *Duties of the vizier*)

van Siclen, Charles C., III. "Ostracon BM 41228: A sketch plan of a shrine reconsidered." *Göttinger Miszellen* 90 (1986): 71–77. (Van Siclen, "Ostracon BM 41228")

Verner, Miroslav. *Abusir: The Realm of Osiris*. Cairo and NewYork: The American University in Cairo Press, 2002. (Verner, *Abusir*)

———. *Forgotten Pharaohs, Lost Pyramids: Abusir*. Prague: Academia Skodaexport, 1994. (Verner, *Forgotten Pharaohs*)

———. *The Pyramids*. New York: Grove Press, 2001. (Verner, *Pyramids*)

———. "*ṯbt*—ein bisher unbekanntes altägyptisches Maß?" *Mitteilungen des Deutschen Archäologischen Instituts Abteilung Kairo* 37 (1981): 479–81. (Verner, "Unbekanntes Maß")

Vernus, Pascal. *Future at Issue. Tense, Mood and Aspect in Middle Egyptian: Studies in Syntax and Semantics* (Yale Egyptological Studies 4). New Haven, CT: Yale Egyptological Seminar, 1990. (Vernus, *Tense, mood and aspect*)

Vernus, Pascal. *Sagesses de l'Egypte pharaonique*. Paris: Imprimerie Nationale, 2001. (Vernus, *Sagesses*)

Vetter, Quido. "Problem 14 of the Moscow mathematical papyrus." *Journal of Egyptian Archaeology* 19 (1933): 16–18. (Vetter, "Problem 14")

Virey, Philippe. *Le tombeau de Rekhmara*. Paris: Leroux, 1889. (Virey, *Rekhmara*)

Vleeming, Sven P. *Papyrus Reinhardt. An Egyptian Land List from the Tenth Century BC*. Berlin: Akademie Verlag, 1993. (Vleeming, *Papyrus Reinhardt*)

Vogel, Kurt. *Die Grundlagen der ägyptischen Arithmetik in ihrem Zusammenhang mit der 2:n-Tabelle des Papyrus Rhind*. Dissertation München, 1929. Reprint Vaduz (Liechtenstein): Saendig Reprint, 1970. (Vogel, *Grundlagen der ägyptischen Arithmetik*)

———. "The truncated pyramid in Egyptian mathematics." *Journal of Egyptian Archaeology* 16 (1930): 242–49. (Vogel, "Truncated pyramid")

van der Waerden, Bartel L. " The (2:*n*) table in the Rhind papyrus." *Centaurus* 23 (1980): 259–74. (van der Waerden, "(2:*n*) table ")

Vymazalova, Hana. "The ʿḥʿ-problems in ancient Egyptian mathematical texts." *Archiv Orientalni* 69 (2001): 571–82. (Vymazalova, "ʿḥʿ-problems")

———. *Staroegyptská matematika. Hieratické matematické texty*. Prague: Český egyptologický ústav, 2006. (Vymazalova, *Hieratické matematické text*)

———. "The wooden tablets from Cairo: The use of the grain unit ḥḳꜣt in ancient Egypt," *Archiv Orientalni* 70 (2002): 27–42. (Vymazalova, "Wooden tablets")

Warburton, David A. *State and Economy in Ancient Egypt: Fiscal Vocabulary of the New Kingdom* (Orbus Biblicus et Orientalis 151). Fribourg and Göttingen: University Press Fribourg Switzerland and Vandenhoeck and Ruprecht, 1997. (Warburton, *State and economy*)

Wengrow, David. *The Archaeology of Early Egypt: Social Transformations in North-East Africa, c.10,000 to 2650 BC*. Cambridge: Cambridge University Press, 2006. (Wengrow, *Early Egypt*)

Wild, Henri. *Le tombeau de Ti*. Vols. 2 and 3 (MIFAO 65). Cairo: Institut Français d'Archéologie Orientale, 1953 and 1966. (Wild, *Ti*)

Wilkinson, Toby. *Early Dynastic Egypt*. New York: Routledge, 1999. (Wilkinson, *Early dynastic Egypt*)

———. *Royal Annals of Ancient Egypt. The Palermo Stone and its Associated Fragments*. London and New York: Kegan Paul International, 2000. (Wilkinson, *Palermo Stone*)

———. *State formation in Egypt: Chronology and Society* (BAR international series 651). Oxford: Tempus Reparatum, 1996. (Wilkinson, *State formation*)

Winlock, Herbert E. *Models of Daily Life in Ancient Egypt from the Tomb of Meket-Re at Thebes* (Publications of the Metropolitan Museum of Art, Egyptian Expedition 18). Cambridge, MA: Metropolitan Museum of Art, 1955. (Winlock, *Models of daily life*)

Wirsching, Armin. *Obelisken transportieren und aufrichten in Ägypten und in Rom*. Norderstedt: Books on Demand, 2007. (Wirsching, *Obelisken transportieren*)

Žába, Zbyněk (ed.). *Preliminary Report on Czechoslovak Excavations in the Mastaba of Ptahshepses at Abusir*. Prague: Charles University, 1976. (Žába, *Ptahshepses*)

Ziegler, Christiane. "La princesse Néfertiabet." In Christiane Ziegler, ed. *L'art égyptien au temps des pyramides*. Paris: Réunion des musées nationaux, 1999, 207–8.

Zonhoven, L.M.J. "A systematic bibliography on Deir el- Medina." In R. J. Demarée and J. J. Janssen, eds., *Gleanings from Deir el-Medîna* (Egyptologische Uitgaven 1). Leiden: Nederlands Instituut voor het Nabije Oosten, 1982, 245–298. (Zonhoven, "Bibliography Deir el Medina")

Subject Index

Egyptian Words and Phrases Index

Index of Mathematical Texts

Hieratic

Demotic